International Trade in Agricultural Products

International Trade in Agricultural Products

Michael R. Reed

Prentice Hall

Upper Saddle River, New Jersey 07458

Library of Congress Cataloging-in-Publication Data

Reed, Michael R.
    International trade in agricultural products / Michael R. Reed.-- 1st ed.
      p. cm.
    Includes bibliographical references and index.
    ISBN 0-13-084209-5
    1. Produce trade--Government policy--United States. 2. Agriculture and state--United
States. I. Title.

HD9000.6 .R436 2001
382'.41'0973--dc21   00-020569

Publisher: Charles Stewart, Jr.
Executive Editor: Debbie Yarnell
Associate Editor: Kate Linsner
Assistant Editor: Kimberly Yehle
Production Editor: Lori Harvey, Carlisle Publishers Services
Production Liaison: Eileen O'Sullivan
Director of Manufacturing & Production: Bruce Johnson
Managing Editor: Mary Carnis
Manufacturing Buyer: Ed O'Dougherty
Art Director: Marianne Frasco
Marketing Manager: Chris Bracken
Cover Design Coordinator: Miguel Ortiz
Cover Design: Liz Nemeth
Interior Design: Carlisle Communications, Ltd.
Composition: Carlisle Communications, Ltd.
Printing and Binding: R.R. Donnelley Harrisonburg

Prentice-Hall International (UK) Limited, *London*
Prentice-Hall of Australia Pty. Limited, *Sydney*
Prentice-Hall Canada Inc., *Toronto*
Prentice-Hall Hispanoamericana, S.A., *Mexico*
Prentice-Hall of India Private Limited, *New Delhi*
Prentice-Hall of Japan, Inc., *Tokyo*
Pearson Education Asia Pte. Ltd., *Singapore*
Editora Prentice-Hall do Brasil, Ltda., *Rio de Janeiro*

10 9 8 7 6 5 4 3 2 1
ISBN 0-13-084209-5

Dedicated to my family, especially Gail, Laura, and Brian, and the University of Kentucky's College of Agriculture, which has been tremendously supportive throughout my career.

Contents

CHAPTER 7 PREFERENTIAL TRADE AGREEMENTS 97

CHAPTER 8 MACROECONOMICS AND ITS INFLUENCE ON INTERNATIONAL TRADE 115

CHAPTER 9 TRADE AND THE ENVIRONMENT 132

Preface

International trade is vital to the agricultural sector in the United States. International trade has allowed U.S. productive capacity in agriculture to expand without seriously eroding prices, and there is no question that trade will become even more critical to U.S. agriculture in the future. It is surprising, then, that there is no textbook at the undergraduate or graduate level that provides a complete treatment of the major issues in international agricultural trade. This book is an effort to provide that complete treatment.

This book covers all the essential topics in an agricultural trade policy course: gains from trade, agricultural trade policies (of exporters and importers), exchange rates, and multilateral trade negotiations. These have been key elements in agricultural trade classes for the past several decades. These topics are fundamental to understanding how the current trade regime works and which parties benefit and lose as the regime changes.

Yet the book also presents concepts on issues that have only recently become important to a fundamental understanding of agricultural trade: the environment, preferential trade agreements, technical barriers, and flexible exchange rates. Without a clear understanding of these new trends in agricultural trade, one cannot fathom where U.S. agriculture has been or will move in the coming decades.

The final four chapters of the book cover company issues that shed light on what helps firms succeed in international markets. This should help instructors who teach in programs that are more agribusiness oriented. The chapters on foreign direct investment and competitiveness take a large-picture view of factors influencing firm behavior and success, while the chapters on export analysis and strategy are oriented toward steps that firms must take in entering and expanding their international markets.

The list of topics is long and challenging for an agricultural trade text. It is highly unlikely that all of them can be covered adequately in a single semester. Some of these issues are covered in separate courses (international finance or multinational corporate behavior) or books (competitiveness and trade/environment), so some chapters move rapidly. Nonetheless, the structure of the book allows professors to pick and choose among topics of interest. One natural route is to take a pure policy approach by covering the first ten chapters. Another route is to cover the basic policy chapters (1 through 5), then cover firm-level issues (Chapters 11 through 14). Yet, the chapters are compartmentalized so that numerous combinations make sense.

The chapter on European agriculture merits special mention because it is quite different from the others. The European Union is mentioned throughout the book, so it is important to understand the E.U. context. In addition, I assign a term paper in my class that requires the student to perform an in-depth analysis of a particular country—its demographics, overall and agricultural trade pattern, macroeconomic policies, agricultural trade barriers, stance in the GATT negotiations, and involvement in preferential trade agreements—but no student is allowed to choose the European Union. Chapter 10 serves as a prototype for certain elements of the term paper.

This text is written for those who have had an intermediate microeconomics class because trade issues can be understood best through extensive graphical analysis. An introductory agribusiness course is also useful if a firm-oriented approach is taken. I have attempted to make linkages back to everyday life through examples and case studies so that the learning experience is enhanced. Any professor can find numerous current events that will support chapters throughout a typical semester.

I owe thanks to Arief Iswariyai and Linda Inman, who helped with graphics and other technical aspects; and four reviewers who provided helpful comments, including Glenn C. W. Ames, University of Georgia; James G. Beierlein, Penn State University; William M. Park, University of Tennessee; and Tim Woods, University of Kentucky. The remaining errors and omissions are mine, but I hope good friends will help me improve the manuscript in the future.

International Trade in Agricultural Products

Chapter 1

An Introduction to Agricultural Trade

World trade has experienced an extraordinary expansion in the last three decades. Technological changes in transportation and communication, a more open world financial and trading system, and increased incomes in many areas of the world have led to these changes. A number of countries have successfully used the world market as their springboard to economic development. Other countries have seen their economic progress hindered because they discouraged imports and other foreign influences. In the last two decades, most countries have become convinced that they must take advantage of increased globalization in order to chart the optimal path for their domestic economies.

The United States has always had a relatively open economy, but international trade has become increasingly important. In the 1950s, U.S. exports and imports accounted for only 3.5 percent and 3.0 percent of U.S. gross domestic product (GDP), respectively (U.S. Bureau of Census). The domestic market was crucial to the well-being of companies because the U.S. growth rate was higher than in the rest of the world. International trade was something that other countries needed to support their economies; U.S. companies simply had to take care of U.S. consumers to succeed. People knew little about countries outside the United States, and most Americans thought that the United States would continue to be so dominant that what happened internationally would be only mildly relevant to the U.S. economy. The U.S. dollar was fixed relative to the price of gold, so the international market was quite stable and few international disturbances were transmitted into the U.S. economy.

The U.S. trade pattern changed slowly through the 1960s, when exports and imports grew faster than U.S. sales and companies began to see that the international market would play an important role in the future U.S. economy (Table 1.1). However, the big changes came about in the 1970s. The U.S. dollar was devalued twice relative to gold in the early 1970s, and finally the fixed exchange

1

TABLE 1.1 U.S. Merchandise Exports and Imports since 1950, Selected Years, in Billion Dollars

	Exports	Imports
1950	$ 10.3	$ 8.9
1955	$ 15.5	$ 11.3
1960	$ 20.4	$ 15.1
1965	$ 27.2	$ 21.4
1970	$ 42.6	$ 40.0
1975	$107.7	$ 98.5
1980	$216.7	$244.9
1985	$206.9	$345.3
1990	$393.6	$495.3
1995	$584.7	$743.4
1998	$670.2	$917.2

Source: U.S. Bureau of Census.

rate standard, which was established as part of the Bretton Woods Agreement in 1944, was abandoned in 1973, allowing the value of the U.S. dollar to fluctuate on a daily basis.

The freeing of the exchange rate forced U.S. companies to realize that they were now part of a global economy where happenings in the rest of the world would have an impact on the United States. The exchange rate is one of the important devices that translates the world economy into the U.S. economy, and its value has a tremendous effect on U.S. prices, trade patterns, and income levels (Schuh). By 1975, U.S. exports accounted for 6.7 percent of GDP and imports were 6.2 percent (U.S. Bureau of Census).

Some of these increases in trade were the result of changing relative prices of oil and other raw materials during the early 1970s, but most of the impact was simply a more open world trading environment due to the new exchange rate regime, trade liberalization through the General Agreements on Tariffs and Trade (GATT), and the fact that most foreign countries were growing faster than the United States.

With the more open trading environment and the significant economic strides that many countries had made, the United States found that many of its industries faced enhanced competition from foreign companies. U.S. manufacturers of machinery and transportation equipment, which accounted for 35 to 40 percent of U.S. exports in the 1950s and 1960s, found themselves struggling to compete with foreign suppliers in international and domestic markets. Further, imports of machinery and transportation equipment (especially from Europe and Japan), which accounted for only 10 percent of imports in the 1950s and 1960s, suddenly grew to 28 percent of a much larger import volume in the 1970s and 1980s (U.S. Bureau of Census). Other U.S. manufacturers have faced similar competitive problems since the 1960s.

Chapter 1   An Introduction to Agricultural Trade

TABLE 1.2 U.S. Agricultural Exports and Imports since 1950, Selected Years, in Billion Dollars		
	Exports	Imports
1950	$ 3.4	$ 5.1
1955	$ 3.5	$ 4.1
1960	$ 4.5	$ 4.0
1965	$ 6.1	$ 4.0
1970	$ 6.7	$ 5.6
1975	$21.9	$ 9.3
1980	$41.2	$17.4
1985	$29.0	$20.0
1990	$39.4	$22.8
1995	$56.2	$30.1
1997	$57.1	$36.0

Source: Economic Research Service, U.S. Department of Agriculture.

The upward trend in U.S. imports and exports has continued in current dollar terms and in percentage of GDP through 1996, when exports accounted for 8.2 percent of GDP and imports accounted for 10.4 percent. It appears that the trend toward increased trade will continue, though imports as a percentage of GDP may grow at a slower rate or even stay constant in the years ahead. U.S. exports may increase faster as macroeconomic conditions in the United States improve and incomes in other countries grow. The United States has had a merchandise trade deficit since 1976, which has some consequences that will be discussed in a later chapter.

OVERVIEW OF U.S. AGRICULTURAL EXPORTS AND IMPORTS

International trade has always been important for agricultural products, especially U.S. agriculture.[1] Countries import many items that are difficult or impossible to produce given their climate and soils. Countries also import agricultural products that are produced at a lower cost by foreign countries. U.S. farmers have relied heavily on exports to provide markets for many of their products. The productive capacity of U.S. agriculture is simply greater than domestic consumers can absorb at market prices.

U.S. agricultural exports averaged slightly greater than 10 percent of farm income in the 1950s and 1960s (Table 1.2). During that time, exports grew from

[1]The U.S. Department of Agriculture's definition of agricultural products is used throughout this book. Most products that are derived from raw agricultural products are referred to as *agricultural*.

$3.4 billion in 1950 to $6.7 billion in 1970 (Economic Research Service). It was during the 1970s, though, when agricultural exports exploded—from $9.4 billion in 1972 to $21.9 billion in 1974, then on to $34.7 billion in 1979. The world had opened to agricultural trade, and the United States was taking full advantage of that situation.

Agricultural exports from the United States grew rapidly in the 1970s. Agricultural producers in this decade experienced a much lower-valued U.S. dollar, crop shortfalls in 1973 and 1974, the Russians becoming more aggressive in purchasing feed ingredients, two oil price shocks, and other factors that served to increase agricultural prices and trade. It was a great time for U.S. agriculture when production was expanded and U.S. agricultural exports accounted for 15 to 27 percent of farm income (U.S. Department of Agriculture).

U.S. agricultural exports hit a peak of $43.3 billion in 1981, a level that was not surpassed until 1994 (Economic Research Service). The rest of the world caught up with the United States during the 1980s, though. Foreign countries responded to more favorable agricultural prices from the 1970s, and foreign producers increased production greatly. South America and Europe became major crop producers and exporters. The value of the U.S. dollar soared in the early 1980s, cutting into U.S. agricultural exports. Exports reached a nadir of $26.2 billion in 1986, just after the peak in the dollar's value. U.S. agricultural policies were not flexible, so many U.S. crop producers found themselves shut out of the export market because government support prices were above world levels.

The decade of the 1990s has seen a solid, positive trend in U.S. agricultural exports (this trend began in 1987), with U.S. exports reaching an all-time high of $60.3 billion in 1996 (Economic Research Service). Exports normally account for more than 20 percent of farm income, and there is strong evidence that agricultural exports will continue to expand. Yet the future of agriculture in the former Soviet Union and Eastern Europe, further agricultural reforms in the European Union, and the impact of recent U.S. agricultural policy changes (the Federal Agriculture Improvements and Reform Act of 1996) will have tremendous effects on future U.S. agricultural exports.

The pattern of U.S. agricultural imports is much easier to explain (Table 1.2). There is a clear, steady upward trend in agricultural imports since World War II, with imports moving from around $4 billion in the 1950s and 1960s to around $35 billion in the late 1990s. Another important trend is that the percentage of imports that compete with domestic production continues to increase—from 40 to 45 percent in the 1950s and 1960s to 70 to 75 percent today (Economic Research Service). American diets are increasingly moving toward temperate climate products such as meat, vegetables, and fruits. Imports of noncompetitive items such as coffee, tea, and cocoa are growing very slowly or not growing at all.

OVERVIEW OF WORLD AGRICULTURAL TRADE

The value of world agricultural exports has increased by over ten-fold in nominal terms during the last thirty years—from $43 billion in 1966 to $464 billion in 1996

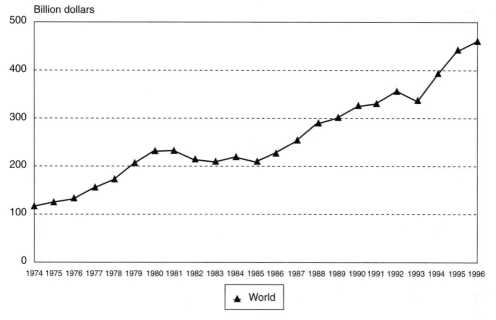

FIGURE 1.1
World agricultural trade, 1974–1996.

(Figure 1.1).[2] Some of that increase has come about through increased prices, but world trade volume has also expanded. Agricultural exports increased modestly in most years, but there were some outliers. World trade jumped by 46 percent between 1972 and 1973, when there was a major crop shortfall, agricultural prices increased, and the Soviet Union purchased much grain to overcome shortages. Before 1972, the growth in world agricultural exports was slow and export values were low.

Since 1974, most years have seen $20 to $50 billion increases in world agricultural exports, but the years 1980–1986 were exceptions to this rule. World agricultural exports reached a peak in 1980 (of $234 billion) which was not surpassed until 1987. These years, 1980 to 1987, coincided with a worldwide economic recession, large crops in many countries, and low agricultural prices. There was also a general increase in nontariff barriers for agricultural products, which reduced trading volumes.[3] These increased trade restrictions and reduced export levels led to the inclusion of agriculture in the Uruguay Round of the GATT, which will be discussed in detail later.

The United States has consistently been the leading agricultural exporter in the world (Figure 1.2). Its exports typically account for 12 to 19 percent of world

[2]All data on world agricultural exports and imports come from the Food and Agriculture Organization, United Nations.
[3]A *tariff* is a tax on a product as it leaves (export tariff) or enters (import tariff) a country. *Nontariff barriers* are policies that impede trade but are not taxes. Examples would include quotas, licensing requirements, and other restrictions on quantity.

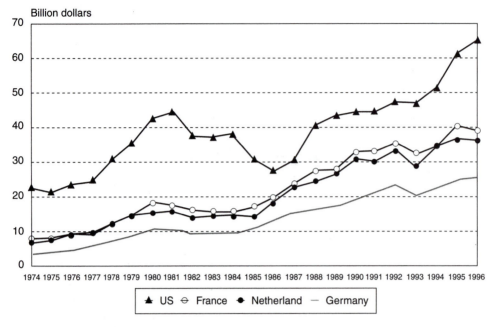

FIGURE 1.2
Agricultural exports of selected countries, 1974–1996.

exports, though there are cyclical patterns. During the mid-1970s through the mid-1980s the United States accounted for 15 to 19 percent of exports, but in most other years it has been somewhat below these levels. U.S. agricultural exports in 1996 totaled $66 billion (an all-time record).[4]

France is currently the second-leading agricultural exporter, followed by the Netherlands and Germany (Table 1.3). A high percentage of the agricultural exports from these countries is intermediate and processed agricultural products that are traded within Europe. The table also shows agricultural exports from the E.U.-15 (European Union) at $56.9 billion. Geographic proximity and free trade among many European countries encourage these exports. There is also much more commonality among European countries concerning consumer tastes, food cultures, and food marketing systems.

Countries that specialize in bulk agricultural products were more important exporters in the 1960s and 1970s. In 1965, Australia was the second-leading agricultural exporter, with Canada and Argentina ranked fifth and sixth, respectively, behind France and the Netherlands. However, the tremendous growth in processed food trade has encouraged export growth from processed food companies in France, the Netherlands, and Germany.

---

[4] This number is from the Food and Agriculture Organization, United Nations. It differs from the U.S. Department of Agriculture number because of differences in definitions for agricultural products.

TABLE 1.3	Leading Agricultural Exporters, in Billion Dollars, 1996
United States	$66.3
France	$40.4
Netherlands	$37.3
Germany	$26.5
Belgium-Luxembourg	$19.0
Italy	$16.9
Australia	$16.1
United Kingdom	$15.4
Spain	$15.0
Canada	$14.7
E.U.-15	$56.9

Source: FAO, United Nations.

TABLE 1.4	Leading Agricultural Importers, in Billion Dollars, 1996
Germany	$44.8
Japan	$41.8
United States	$37.9
France	$27.6
United Kingdom	$26.7
Italy	$25.6
Netherlands	$20.7
China	$17.5
Belgium-Luxembourg	$17.0
Spain	$13.1
E.U.-15	$64.2

Source: FAO, United Nations.

The leading agricultural exporting countries are also leading agricultural importers, which indicates the diversity of agricultural production (Table 1.4). In 1996, Germany was the leading agricultural importer, followed by Japan, the United States, France, the United Kingdom, and Italy. If the E.U.-15 is considered a single unit, it is clearly the leading agricultural importer with $64.2 billion in 1996. The prominence of European countries as importers is further evidence of the importance proximity and free trade have on agricultural trade.

The next section gives a brief introduction to world trade in some of the most important agricultural products. The leading exporters and importers are given for these products, and some trade values are presented. One can obtain more detailed information from the Food and Agriculture Organization (FAO) of the United Nations (UN), which is the source of all data presented in the following section.

Commodity Highlights

World wheat exports are dominated by four countries: the United States, Canada, Australia, and France. These countries exported $6.3 billion, $3.4 billion, $3.1 billion, and $2.8 billion, respectively, in 1996. Argentina is also an important exporter in certain years. Wheat export patterns have been distorted heavily in the past because of subsidies by the United States and European countries. The recent GATT agreement will likely change the pattern of wheat exports significantly over time toward exports from Canada, Australia, and Argentina. The leading wheat importers are China ($2.1 billion in 1996), Japan ($1.6 billion in 1996), Italy, Brazil, and Egypt.

The United States normally exports more corn than the rest of the world combined. In 1996, U.S. corn exports totaled $8.6 billion while world exports totaled $12.7 billion. That year saw very large U.S. exports, but the United States regularly accounts for 60 percent of world exports. France and Argentina are also important corn exporters. Corn importers are more diversified, with Japan being the leading destination (typically accounting for about 20 percent of world imports). China and Korea are also large importers.

World soybean exports are also dominated by the United States, which normally accounts for more than 50 percent of world exports (though other countries are more important exporters of soybean oil and meal). The United States exported $7.5 billion of soybeans in 1996, while Brazil, the second-leading exporter, exported $1.0 billion. Argentina is also a large soybean exporter. Import destinations are more diversified, with Japan being the largest importer each year (importing $1.7 billion in 1996). Other important soybean-importing countries are the Netherlands, China, Mexico, Germany, and Spain.

The United States is the world's largest exporter of beef and poultry, exporting $2.4 billion and $2.5 billion in 1996, respectively. Australian beef exports are often quite close to U.S. levels, while U.S. poultry exports only recently overtook French poultry exports for the number one spot. Other important beef exporters are the Netherlands, France, Ireland, and Germany. The Netherlands, Brazil, and China are also important poultry exporters. Despite rapid increases in world poultry trade in recent years (it has increased 75 percent between 1993 and 1996), it was still below the level of world beef trade in 1996 ($14.5 billion for beef versus $10.8 billion for poultry).

Japan is the leading destination for beef and poultry shipments, importing $2.8 billion of beef in 1996 and $1.6 billion in poultry. Japanese beef imports have grown rapidly since 1988, when their market was liberalized, and they now account for 20 to 25 percent of world beef imports. Other important beef importers are the United States and Germany. Poultry destinations are more dispersed, with Germany's import level often close to Japan's. Other important poultry-importing countries are the United Kingdom, Hong Kong, and the former Soviet Union.

More pork is consumed worldwide than any other meat, and it is also the meat that is most traded: almost $17.0 billion was imported in 1996. This is due to the great tradition of pork consumption in Europe and Asia. The leading pork-exporting country is Denmark, with exports of $3.3 billion in 1996, followed by the

TABLE 1.5	U.S. Agricultural Exports by Major Products, 1997, in Billion Dollars	
Soybeans		$ 7.38
Coarse grains		$ 5.98
Red meat		$ 4.49
Wheat		$ 4.10
Cotton		$ 2.71
Poultry meat		$ 2.42
Fresh fruit		$ 2.10
Processed fruits and vegetables		$ 2.09
Total		$57.14

Source: Economic Research Service, U.S. Department of Agriculture.

Netherlands, China, and Belgium-Luxembourg. The leading pork-importing countries are Japan ($4.1 billion in 1996), Germany ($2.5 billion in 1996), Italy, and the United Kingdom.

U.S. AGRICULTURAL TRADE BY COUNTRY AND COMMODITY

As seen earlier, the United States is the leading agricultural exporting country in the world. Table 1.5 gives a product breakdown of U.S. agricultural exports for 1997. The rankings for these products are sensitive to the product definitions, but the table gives a good overview.

Bulk commodities—such as soybeans, coarse grains, wheat, and cotton— dominated U.S. agricultural exports in 1997. In fact, bulk commodities have historically dominated U.S. agricultural exports, but there has been a trend toward increased exportation of intermediate and consumer-oriented agricultural products in recent years. Recently, exports of red meat, poultry, fruits, and vegetables have grown more rapidly than bulk products.

Figure 1.3 shows U.S. exports of bulk, intermediate, and consumer-oriented products since 1970. Note the steady increase in the last two categories, while the bulk category fluctuates greatly on a yearly basis. Exports of bulk commodities are more price-sensitive, while markets for intermediate and consumer-oriented products are less susceptible to price competition. In the mid-1970s (1973 to 1976), bulk exports accounted for over 70 percent of U.S. agricultural exports. In the mid-1980s, bulk exports shrank to less than 55 percent of U.S. agricultural exports because of pricing pressures and the high value of the U.S. dollar. The early 1990s also saw the bulk export percentage shrink quickly to less than 45 percent of agricultural exports as a result of pricing pressures from a worldwide recession. The percentage of bulk exports reached a low in 1997, when they accounted for 41 percent of agricultural exports.

U.S. Department of Agriculture's Product Classifications

The U.S. Department of Agriculture classifies agricultural products into various categories. Bulk commodities have a value of less than $400 per metric ton. Intermediate products have been processed in some way, but are not ready for ultimate consumption. Consumer-oriented products are ready for ultimate consumption (they may have been processed in some way). At times the term *manufactured* or *processed food* is used in this book. This refers to any agricultural product that has been materially transformed in some way. It includes intermediate and some consumer-oriented products.

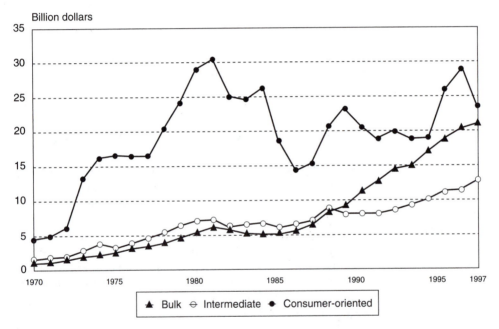

FIGURE 1.3
United States agricultural exports of bulk, intermediate, and consumer-oriented products, 1970–1997.

These U.S. agricultural exports reach almost every country of the world, but as Table 1.6 indicates, nearly 40 percent of them go to Japan, Canada, and Mexico. The Japanese have a large high-income population on a crowded, mountainous island where agricultural production potential is limited. It has been the largest importer of U.S. agricultural products for many years. Exports to Canada received a big boost when the United States and Canada signed a free trade agreement, which took effect in January 1989. Exports to Mexico are large because of Mexico's proximity, large population, and poor growing conditions for many agricultural prod-

TABLE 1.6	Leading Destinations for U.S. Agricultural Exports, 1997, in Billion Dollars	
Japan		$10.52
Canada		$ 6.77
Mexico		$ 5.17
Korea		$ 2.85
Taiwan		$ 2.61
Netherlands		$ 1.92
Hong Kong		$ 1.71
Peoples Republic of China		$ 1.60
Germany		$ 1.32
United Kingdom		$ 1.31
E.U.-15		$ 8.90
Total		$57.14

Source: Economic Research Service, U.S. Department of Agriculture.

TABLE 1.7	U.S. Agricultural Imports by Major Products, 1997, in Billion Dollars	
Raw coffee		$ 3.21
Wine and beer		$ 3.20
Fresh fruits		$ 2.91
Red meat		$ 2.63
Processed fruits and vegetables		$ 1.96
Fresh vegetables		$ 1.72
Snack foods		$ 1.67
Live animals		$ 1.60
Total		$35.97

Source: Economic Research Service, U.S. Department of Agriculture.

ucts: as will be seen later, the free trade agreement with Mexico, which took effect in January 1994, has had a minor impact on U.S. agricultural exports thus far. Other important importing countries are in East Asia (Korea, Taiwan, Hong Kong, and China) and Europe (the Netherlands, Germany, and the United Kingdom).

The United States is also the fourth largest importer of agricultural products. Table 1.7 gives a product breakdown of U.S. agricultural imports for 1997. Again, the rankings for these products are sensitive to the product definitions. Intermediate and consumer-oriented products account for a much larger percentage of U.S. agricultural imports than exports. Wine, beer, fresh fruits, meats, vegetables, and other valued products account for 76 percent of U.S. agricultural imports, and many of the bulk agricultural products that are imported are not produced in significant amounts within the United States (coffee, rubber, cocoa,

TABLE 1.8	Leading Origins for U.S. Agricultural Imports, 1997, in Billion Dollars
Canada	$ 7.39
Mexico	$ 4.08
Indonesia	$ 1.57
Brazil	$ 1.47
Colombia	$ 1.43
Italy	$ 1.37
Netherlands	$ 1.25
France	$ 1.23
Australia	$ 0.96
Thailand	$ 0.85
E.U.-15	$ 6.95
Total	$35.97

Source: Economic Research Service, U.S. Department of Agriculture.

and tea). Consumer-oriented products account for 57 percent of U.S. agricultural imports, and these imports increase consistently as U.S. income grows, accounting for the steady increase in total U.S. agricultural imports over time (Table 1.2).

U.S. agricultural imports from Canada have been growing rapidly in recent years, and most of those imports are processed food items (Table 1.8). The U.S. free trade agreement with Canada has also stimulated Canadian agricultural exports such as red meat, live animals, vegetables, and grains. Mexico exports to the United States large quantities of vegetables, coffee, fruits, and beer. Other countries have climates that enable them to export many items that cannot be produced competitively on a large scale in the United States. From Indonesia, the United States imports mostly rubber, cocoa, and coffee; from Brazil, it imports coffee, tobacco, nuts, and orange juice; and from Colombia, it imports coffee, bananas, and nursery products.

IMPORTANT ISSUES IN AGRICULTURAL TRADE

Barriers to international trade and their impacts have always been important issues for agriculture, and their importance has heightened over the years. Much of this importance stems from the trend toward nontariff barriers for agricultural products, which had become increasingly prevalent during the 1980s. The Uruguay Round of the GATT was very successful in reversing this trend toward increased restrictions on agricultural trade, and the new World Trade Organization (WTO) has rules that will help preserve this freer trade environment. Nonetheless, freer trade in agriculture continues to be an important issue.

Three chapters of this book (Chapters 2, 3, and 4) present the framework to analyze the gains from trade and impacts of import and export barriers. These chapters provide the basis for the rest of the text. Chapter 6 deals specifically with the GATT and WTO, paying particular attention to the recent Uruguay Round and its potential consequences for world trade.

With the recent GATT agreement prohibiting nontariff barriers, there is speculation that technical barriers to agricultural trade will increase. These technical barriers involve sanitary and phytosanitary regulations and other rules for quality, packaging, and labeling. Often these regulations are enacted to enhance food safety or to increase information flows to consumers, but some people argue that these barriers are often intended simply as hurdles to discourage imports. There are new rules and operating procedures to distinguish those technical barriers that have a true scientific or economic rationale from those that do not. Disputes associated with technical barriers will be crucial in the years ahead as standard tariff and nontariff barriers fall or are eliminated. Chapter 5 deals with the issues of technical barriers to trade.

Despite recent multilateral successes in reducing trade barriers, many new free trade zones have been established during the 1980s and 1990s. The first major free trade area, the European Common Market, has now expanded into an economic union that covers most of Western Europe and includes free movement of labor and a common currency. For the United States, a major new era in agricultural trade came about when the North American Free Trade Area (NAFTA) was established in 1994. There are many trade issues that spring from these free trade areas, and they are covered in Chapter 7, with a special reference to NAFTA and what it means to U.S. agriculture.

Changes in the world financial system have had tremendous impacts on agricultural trade. The change from a fixed exchange rate system to a flexible system in 1973 and the concomitant increase in currency flows among countries helped stimulate heightened agricultural trade flows. World financial markets continue to expand exponentially, and an understanding of the relationship between exchange rates, interest rates, and trade is essential in today's increasingly integrated world economy. There are entire books that cover these topics (and other topics in international finance) in great detail. Chapter 8 of this text covers the basics of these world macroeconomic phenomena and their effects on agricultural trade.

The environment and its relation to trade and trade policy is an emerging topic of great significance. Chapter 9 covers environmental issues. As trade barriers fall throughout the world, differences in environmental standards may play an enhanced role in determining the production and trade pattern among countries. Chapter 9 covers conceptual frameworks that incorporate environmental standards in trade models. It also explains how these environmental issues are covered within the new WTO and other international organizations whose role is to increase cooperation among nations.

Chapter 10 provides a case study of European Union (E.U.) agriculture. This case study is chosen because the United States and Europe are embroiled in so many agricultural trade disputes. It is worthwhile to study the background of European agriculture to understand their stance on these issues. The E.U. has also been instrumental in establishing trade barriers that are used by other countries; so again, familiarity with their policy dynamic is fundamental to understanding the international debates on agricultural policy.

Agricultural trade has expanded greatly over the last three decades, but sales of multinational food corporations (food firms that own facilities in more than one country) have expanded even faster. In today's world, one must understand the operations of these multinational food firms in order to comprehend the present and future trends in agricultural trade. Agricultural exporters, especially food processors, increasingly find their products competing in international markets with large-scale companies such as Nestlé, Sara Lee, Unilever, and others that process local agricultural ingredients in their facilities throughout the world. Chapter 11 covers the operations of multinational firms in general and multinational food firms in particular.

Discussion of the operations of multinational firms leads naturally into a general discussion of *competitiveness,* which has been an important "buzzword" in the business and political arenas in recent years. Businesses are seeing their sales erode quickly as dynamic markets change rapidly because of technology, tastes, and regulations. Governments are struggling with policies to increase the competitiveness of their industries. Chapter 12 covers the basic concepts of competitiveness and how they relate to the agricultural and food trade.

The last two chapters of the book (13 and 14), cover export market development. The analyses needed for a reasoned assessment of a firm's capability for competing in export markets are discussed in Chapter 13. The development of a strategy to carry out the export marketing plan is covered in Chapter 14. These firm-level considerations in international trade are vitally important for success.

SUMMARY

1. International trade has become increasingly important to U.S. agriculture since World War II. In the 1950s, exports accounted for 3.0 percent of GDP, while in 1996 exports accounted for 8.2 percent. Imports accounted for an even larger percentage of GDP in both years.
2. Trade liberalization, through the GATT, has provided an important stimulus to international trade in all goods. The move to a floating exchange rate system also coincided with a large increase in U.S. agricultural exports.
3. U.S. agricultural exports grew rapidly in the 1970s, but they stagnated throughout the 1980s. Agricultural export growth reemerged in the 1990s, and U.S. agricultural exports hit a peak of $60.3 billion in 1996. U.S. agricultural imports have increased steadily since World War II.
4. The United States is the leading agricultural exporter in the world. Other important agricultural exporters (in order) are France, the Netherlands, and Germany. Germany is the leading agricultural importer in the world, followed by Japan, the United States, and France.
5. U.S. exports of intermediate and consumer-oriented agricultural products have grown as a percentage of total agricultural exports since 1970. In 1997, exports of intermediate agricultural products totaled $12.6 billion, while exports of consumer-oriented agricultural products totaled $21.0 billion.
6. The United States exports agricultural products to most countries of the world, but 40 percent of U.S. agricultural exports go to Japan, Canada, and Mexico.

QUESTIONS

1. What factors have caused U.S. exports and imports to increase so rapidly relative to overall economic activity? Will these factors continue in the future?
2. Will U.S. agriculture continue to rely on exports to fuel farm income growth? Will our agricultural exports become more concentrated in certain products in the future? Will imports become more diverse?
3. What will the U.S. agricultural export pattern be in 2035, and how will that pattern affect U.S. agriculture?
4. In your opinion, what is the most important agricultural trade issue? Why?

REFERENCES

Economic Research Service. U.S. Department of Agriculture. *Foreign Agricultural Trade of the United States.* Washington, DC: Government Printing Office. Web site: *www.fas.usda.gov*

Food and Agriculture Organization (FAO). United Nations. *FAO Trade Yearbook.* New York: Statistical Office. Web site: *www.fao.org*

Schuh, G. Edward. "Exchange Rate and U.S. Agriculture." *American Journal of Agricultural Economics* 56 (1974): 1–13.

U.S. Bureau of Census. *Statistical Abstract of the United States.* Washington, DC: Government Printing Office. Web site: *www.census.gov*

U.S. Department of Agriculture. *Agricultural Statistics.* Washington, DC: Government Printing Office. Web site: *www.usda.gov*

Chapter 2

Gains from Trade

Notice that the title of this chapter is "gains from trade." It is not titled "gains from exports," and it doesn't involve "losses from imports." Countries gain from exporting and importing. This may not be well known in the popular press, but this result is crucial for practicing economists to understand. Trade—exporting and importing—is mutually beneficial for all countries. The gain from trade is one of the fundamental propositions that all economists agree upon, so it is important that students of the subject understand the concepts behind this proposition.

This chapter begins with the concept of comparative advantage and gains from trade in a simple, two-country, two-good world. This is where the analysis begins in texts in international economics. Yet most of the chapter deals with one good and two countries because this is the model that is more commonly used in agricultural trade analysis. The reason that the one-good case is more commonly used than the two-good case is that agricultural economists are often interested in the effects of trade policies on trade flows, which is most easily analyzed in a one-good case.

SIMPLE COMPARATIVE ADVANTAGE

Introductory analysis of gains from international trade normally begins with the concept of comparative advantage, which dates back to British economist David Ricardo (1772–1823). The comparative advantage concept highlights the proposition that relative productivity between countries is more important than absolute productivity in determining trade patterns. The production possibilities frontier (PPF) is an important concept for comparative advantage and gains from trade.

Consider the trade pattern between the United States and France, assuming there are two goods in the economy, bread and wine. Both goods require similar inputs, land, labor, capital, and so on, that are assumed in fixed supply in each country. The production functions for the United States are such that if all inputs

TABLE 2.1	Production Possibilities Data for France and the United States		
U.S.		France	
Bread	Wine	Bread	Wine
120	0	40	0
80	10	20	10
40	20	10	15
0	30	0	20

are used for production of bread, 120 units of bread will be produced; whereas 40 units of bread would be produced if all inputs are used for bread production in France. If the United States and France used all of their resources in the production of wine, they would produce 30 units and 20 units, respectively. Table 2.1 gives the range of production possibilities available to each country, given their available resources.

Notice that the United States can produce more bread or wine than France because it has more abundant resources. The United States is said to have an *absolute advantage* in production of bread and wine. Absolute advantage is irrelevant for international trade because it only gives an indication of the country's size or total resource base. Despite the fact that the United States has an absolute advantage for each good, both countries can gain from trade because their marginal rates of product transformation (the rate at which one product must be sacrificed to produce another) vary for the two commodities. In the United States one must give up 4 units of bread for every unit of wine, while in France one must give up 2 units of bread for every unit of wine. This gives France a *comparative advantage* in production of wine because its resource base is such that its opportunity costs (in terms of bread) are lower for wine than in the United States. The opportunity cost for bread is lower in the United States. These different opportunity costs by country bring about the gains from trade.

If there is no trade between France and the United States, each country must consume what it produces. Thus, the domestic price ratios for bread and wine, which give the signals for what should be produced and consumed, must be consistent with the PPFs (assuming that some of each good is consumed). Without trade, the price of wine must be 4 times the price of bread in the United States and 2 times the price of bread in France; wine is relatively expensive in the United States. If wine cost $1 in the United States, bread would cost $0.25; if wine cost $1 in France, bread would cost $0.50.

The production possibilities frontiers associated with Table 2.1 make it easier to see comparative advantage. Figure 2.1 shows the PPFs for each country. The PPF for the United States lies upward and to the right of the PPF for France, indicating that the United States has an absolute advantage in production of bread and wine. Yet, because the slope of the PPF for the United States is different than the slope for

FIGURE 2.1
Linear production possibilities frontiers
for France and the United States.

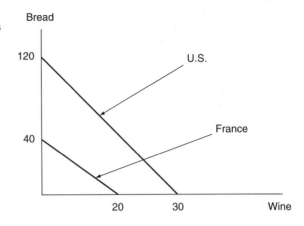

France, welfare-enhancing trade can take place. Without trade, each economy is forced to consume somewhere along its PPF. Thus, the PPF defines what the price ratios will be between bread and wine.

Suppose trade between the United States and France is allowed and the price for wine is 3 times the price of bread in both countries (between the price ratios for the United States and France). Profit-maximizing producers in France would choose to produce only wine, while profit-maximizing producers in the United States would choose to produce only bread. Because trade is possible, French wine producers can trade some of their wine for bread made in the United States (and vice versa). This trade occurs at the international price ratio (or international terms of trade) of 3:1.

The international terms of trade help define consumption possibilities. These consumption possibilities lines define combinations of the two goods that the country can choose to consume based on its production level and the international terms of trade. Figure 2.2 shows the consumption possibilities for each country. Both countries clearly gain because their consumption possibilities lines lie above their PPFs, allowing each country to consume more of each good than when trade was not allowed.

Any wine/bread price ratio between 4 and 2 will result in France specializing in wine production and the United States specializing in bread production. However, the equilibrium price has ramifications on national income. If the international price of wine was 3.99 times the price of bread, incomes would be higher in France, and a price of wine that was 2.01 times the price of bread would make incomes higher in the United States. The consumption possibilities lines associated with those price ratios would reflect those income levels (and the ability to consume the two goods). It would be very steep with a price ratio (price of wine, P_W, divided by the price of bread, P_B) of 3.99:1 and more flat with a price ratio of 2.01:1.

The gains from trade in this analysis can be measured in terms of utility or added consumption of the two goods, as shown in Figure 2.3. Assume that the new terms of trade allow the United States to consume 20 units of wine and 60 units of bread (point B), while the terms of trade allow France to consume 10 units of wine

FIGURE 2.2
Consumption possibilities for France
and the United States.

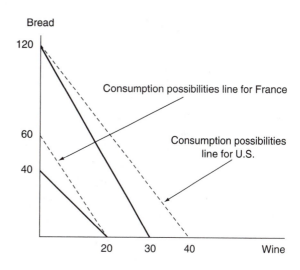

FIGURE 2.3
Consumption and gains given a 3:1
price ratio.

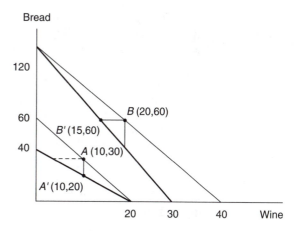

and 30 units of bread (point *A*). The net welfare gain to the United States can be measured in terms of wine by moving back to the United States' original PPF and finding that the United States can consume 5 additional units of wine, while leaving its bread consumption at 60 units. The net welfare gain from trade to France can be measured in terms of bread by going back to France's original PPF and finding that France can consume 10 additional units of bread, while leaving its wine consumption at 10 units. Welfare gains can be measured in units of either good.

GAINS FROM TRADE, TWO GOODS

The simple comparative advantage analysis can provide some insights, but the generalized PPF (with diminishing marginal rates of technical substitution) gives a more realistic view of the gains from trade in a two-good world. Figure 2.4 shows

Production Possibilities Frontiers

Production possibilities frontiers (PPFs), which are derived from production functions for goods, trace the maximum output that can be obtained from an economy assuming that all resources are fully employed. They are analogous to the country's budget constraint if there is no trade, but one must remember that they are in quantity space (the axes are quantities of the two goods). Movement along the PPF shows the effect of moving inputs from production of one good to another. They are usually concave to the origin, as shown in Figure 2.4, because the economy experiences diminishing marginal rates of technical transformation. This technical term means that as more resources are put into production of a good, the marginal productivity of those resources declines; thus an economy is getting less output per unit of input as more input is used.

Not all inputs are equally suited for production of wheat (bread) and grapes (wine), especially when land and labor are involved. Consider the point in Figure 2.4 where the United States produces 120 units of bread and no wine. If the country was going to begin producing wine, it would use the inputs (especially land and labor) that are best suited for wine production for those first units of wine production—generating a relatively large amount of wine with little bread production forgone. As more inputs are put into wine production, those inputs will be less suited to wine production than the first inputs, so less wine will be forthcoming.

Figure 2.1 shows a relationship where all inputs are equally productive in wine and bread production (because the PPF is linear). However, if the inputs that are most suited to wine production are taken out of bread production first, the PPF will lie above the straight line depicted in Figure 2.1 (as shown in Figure 2.4).

a concave PPF (and social indifference curve) for the United States with the two goods, bread and wine. It is assumed that there is perfect competition, no tariffs or taxes, no transportation costs, and the country is small.

When there is no trade, the equilibrium production (and consumption) point is at A in Figure 2.4, corresponding to Q_B units of bread and Q_W units of wine, where the indifference curve UU is tangent to the PPF. This equilibrium is reached because relative prices in the economy are consistent with the line segment PP, which has a slope of $-P_W/P_B$. This relative price gives producers the signal to use their resources to obtain outputs Q_B and Q_W. This relative price also gives the consumers the signal on relative scarcity, which generates the same point for consumption. Given the income constraint that the United States faces (as shown by the PPF), there is no indifference curve that lies further upward and to the right, so welfare in the United States is maximized at Q_B and Q_W.

Indifference Curves

Indifference curves trace consumption points where consumers are indifferent about the consumption bundle. They show the willingness of consumers to substitute one good for another, so indifference curves reflect tastes and preferences of consumers. Indifference curves are always convex to the origin (as shown in Figure 2.4) because of the law of diminishing marginal rate of substitution. This law says that as one consumes more of a product, the marginal utility from consuming an additional unit falls.

There are an infinite number of indifference curves, each depicting a different level of utility. As one moves upward and to the right in Figure 2.4, the utility associated with the indifference curve increases. Society's goal is to get to the indifference curve with the highest utility (or the one that lies the furthest upward and to the right).

FIGURE 2.4
U.S. production and consumption with typical PPFs.

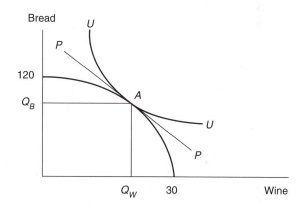

If trade is introduced and there are no barriers to trade, the constraint that the United States consume along its PPF is relaxed because imports and exports are allowed. The *PP* line segment is no longer relevant because the United States can trade at the world relative price, which is likely different than *PP*. If the United States is a small country, meaning that it can buy or sell any amount of a good without affecting the world price (an assumption that will be relaxed later), and the world relative price (or terms of trade) for bread is higher than the U.S. relative price without trade, the United States will export bread and import wine. Figure 2.5 shows this graphically. *P'P'* is the consumption possibilities line.

The world terms of trade (shown as *P'P'*) are the key to the equilibrium with trade. Producers in the United States react to the new higher relative price of bread and increase their production (to Q_{BT}). The only way they can produce more bread is to reduce their production of wine (which falls to Q_{WT}). It is clear, though, that the new

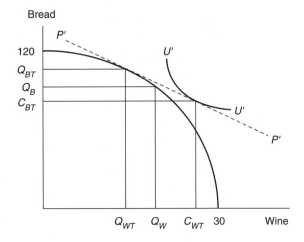

FIGURE 2.5
Production and consumption with a concave PPF.

production point (Q_{BT} and Q_{WT}) generates the maximum income for producers given the new terms of trade. Note that the United States does not specialize in production of bread because it has some resources that are well suited for wine production.

Consumers will react to the new higher relative price of bread, and they will choose to increase their consumption of wine and reduce their consumption of bread. They will maximize their welfare by consuming at C_{BT} for bread and C_{WT} for wine, where the new income constraint along $P'P'$ is tangent to the indifference curve $U'U'$. This is clearly at a higher welfare level than before trade.

The production and consumption points must be on the world terms of trade segment to ensure that there is balance of trade equilibrium (the value of imports is exactly equal to the value of exports). The world terms of trade line is the new budget constraint for the country, not the PPF. The world allows the United States to trade at the new terms of trade as long as the U.S. value of exports equals the value of imports. The world relative price reflects relative scarcity between bread and wine for the world. Because the United States is more productive relative to the rest of the world in bread production, opening up trade gives signals to the United States that it should produce more bread (and consume less bread).

The gains from trade can be measured a number of ways. The first is simply to measure the difference in welfare between the two consumption points (without trade and with trade)—that is, measuring how far $U'U'$ is beyond UU (the point where an indifference curve is tangent to the PPF). However, economists are not very good at assigning welfare values to the abstract notion of an indifference curve. Another way is shown in Figure 2.6 by measuring increased consumption of wine (or bread). In this figure, the consumption point after trade is opened is A. The gain from trade in terms of wine, relative to what could be consumed without trade, is $A - B$, whereas the gain from trade in terms of bread is $A - C$.[1]

---

[1]These distances, $A - B$ and $A - C$, overestimate the gains from trade because they do not consider consumption adjustments that would occur when relative prices change. However, the distances give a concrete measure of gains from trade.

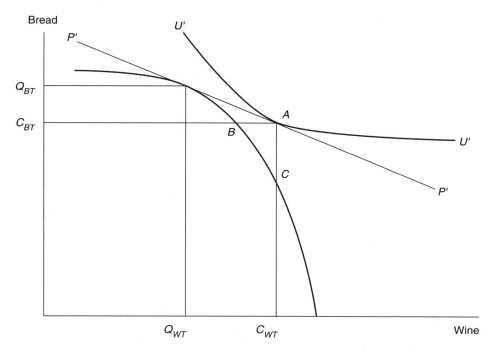

FIGURE 2.6
Gains from trade to the United States.

The Embargo on Iraq

In August 1990, the United Nations imposed a trade embargo on Iraq. No member of the United Nations (UN) could export to or import from Iraq because of Iraq's invasion of Kuwait and its failure to comply with UN weapons inspectors, who were trying to determine whether Iraq was storing or producing biological or chemical weapons. This was not a purely political matter, because the United Nations was using gains from trade (or the lack of gains because of the trade embargo) in order to encourage Iraq to comply.

In April 1995, the UN passed a resolution that allowed Iraq to sell up to $2 billion in oil for food, medicine, and other humanitarian needs every six months. Nonetheless, there is no question that the trade embargo has hurt the economic well-being of the Iraqi people. This unfortunate example, though, is evidence that world leaders realize that countries gain from trade.

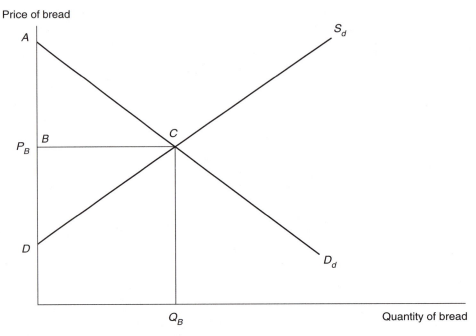

Price of bread

Quantity of bread

FIGURE 2.7
No-trade equilibrium for bread.

PARTIAL EQUILIBRIUM, ONE GOOD

Most international economics books focus on the two-good model and its ramifications on factor markets (such as labor and capital) because they are interested in the effects of trade on factor returns (wages and returns to capital). Agricultural economists tend to focus on output markets and investigate international trade issues as they impact the pattern of trade. Often this analysis of trade patterns is easier to envision in a partial equilibrium model (supply and demand analysis) with one good and many countries. The equilibrium is partial because it includes only the price for the good in question and holds all other good prices constant. This model is an excellent vehicle to show how changes in variables affect the equilibrium for the good in question.

Figure 2.7 shows typical supply and demand curves for bread for a particular country, S_d and D_d, where the d subscript denotes domestic. If there is no trade in bread, then the domestic price (P_B) is where domestic supply equals domestic demand (Q_B). P_B is called the autarkic or no-trade price. At P_B and Q_B producer surplus is the area BCD, consumer surplus is the area ABC, and their sum is maximized (as shown in Figure 2.7). In perfect competition, the supply of bread is the industry marginal cost curve and the demand for bread is the marginal benefit of

Consumer and Producer Surplus

Consumer and producer surplus are common means of measuring welfare, especially in comparative static situations. Consumer surplus measures the welfare gain that consumers obtain from purchasing goods at prices that are below their worth. Given a typical (downward-sloping) individual demand function, the consumer is often willing to pay more for a product than the market price. This net gain that consumers accrue is called a *surplus*. The area below the demand curve and above the market price is called *consumer surplus*. That area is the sum of gains that consumers obtain but do not pay for through the market price.

In a similar manner, producer surplus measures the welfare gain that producers obtain from selling goods at prices that are above their production cost. Given a typical (upward-sloping) marginal cost curve, the producer is often able to produce some units for less than the market price. This net gain that producers accrue is called a *surplus*. The area above the marginal cost curve and below the market price is called *producer surplus*.

bread, so P_B is where the marginal cost of producers equals the marginal benefit to consumers. Algebraically,[2]

$$S_B = f(P_B/\dots)\tag{2.1}$$
$$D_B = g(P_B/\dots)\tag{2.2}$$
$$Q_B = \min(S_B, D_B)\tag{2.3}$$

when surplus is maximized $Q_B = S_B = D_B$ (2.4)

If price is above the intersection of supply and demand (as in Figure 2.8 at P_B'), then producer surplus is area *BCFG*, which could be higher or lower than at the intersection of supply and demand, but consumer surplus is much lower (area *ABC*) and the sum of producer and consumer surplus is less than when domestic supply equals domestic demand. In this case, bread's price does not reflect the true opportunity costs of bread and there is a net loss to society (the triangular area *CEF* in Figure 2.8). The signals are not optimal because consumers are receiving price signals that bread is worth P_B', when it only costs P_B'' to produce. At a price of P_B', consumers will choose to consume only Q_B' units of bread, less than what they would consume if they were given the correct price signal (P_B). A similar story

[2]The slash in the equation means other things held constant. In this example the supply function holds other output prices, all input prices, and other factors that affect marginal cost constant. The demand function holds other good prices, income, and other demand factors constant.

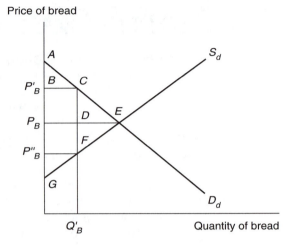

FIGURE 2.8
Consequences of nonequilibrium prices on welfare.

Price of bread

A

S_d

B C

P'_B

P_B D E

F

P''_B

G

D_d

Q'_B Quantity of bread

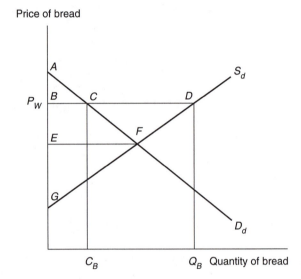

FIGURE 2.9
Gains from trade for the exporting country.

Price of bread

A

S_d

P_W B C D

E F

G

D_d

C_B Q_B Quantity of bread

could be told if bread's price was below the intersection of supply and demand, such as at P_B'', but in this case consumer surplus would be larger than optimal, producer surplus would be smaller, and there would be another net loss to society.

When trade is allowed, though, the constraint that domestic supply equals domestic demand is relaxed. The country can produce a surplus or consume more than it produces as long as international trade makes up the difference. Note, however, that if a country imports one product, it must export at least one product to preserve its trade balance. Figure 2.9 shows an example where a small country becomes an exporter of bread, since it produces a surplus at the world price that it faces. If the world price of bread (P_w) is above the autarkic price, then the country will export. Domestic production will be Q_B, domestic consumption will be C_B, and

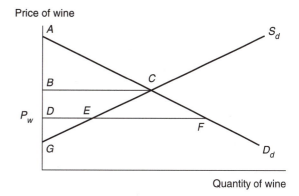

FIGURE 2.10
Gains from trade for the importing country.

Price of wine

P_w

Quantity of wine

the difference will be exported. Producer surplus in Figure 2.9 is *BDG*, consumer surplus is *ABC*, and the net gain to the country from trade is the triangular area *CDF*. Notice that relative to no-trade, consumers clearly lose (area *BCFE*), while producers clearly gain (*BDFE*).

If trade is opened up, bread producers who gain could compensate consumers fully for their loss from the higher price of bread and there would still be net gains to bread producers. This doesn't normally happen because there are many goods that are traded and the country will import some of those goods, making consumers net gainers in total. Figure 2.10 shows that consumers clearly gain when trade is opened and the good is imported. If the United States allowed wine imports at the world price (P_w), consumers would gain *BCFD* from the lower price of wine, but wine producers would lose *BCED*. In total, though, the country would have a net gain of *CEF* resulting from trade in wine.

Figures 2.9 and 2.10 allow the gains from trade shown in Figures 2.5 and 2.6 to be seen on a single-good basis. The gains are still conceptual, especially the gains to consumers, but they are more measurable than the pure utility concepts used in Figure 2.5.

A critically important concept throughout the rest of this book is effective supply and demand. Effective supply and demand measure the impact of domestic and world markets on the country. Figure 2.11 shows that the effective demand curve for a small bread-exporting country is perfectly elastic at the world price, P_w, meaning that the country can export any amount that it wants at the world price. The effective demand curve, D_e, is the sum of the domestic demand curve (with its normal downward slope) and the international demand curve, which is perfectly elastic at the world price. The domestic price, P_d, assuming there are no trade barriers, is determined by the intersection of the supply curve and the effective demand curve.[3]

Figure 2.12 shows that the effective supply curve (S_e) for a small wine-importing country, which is the sum of the upward-sloping domestic supply curve

[3]One can think of the domestic supply curve as the effective supply curve for the exporting country. In a similar manner, the domestic demand curve is the effective demand curve for the importing country.

FIGURE 2.11
Effective demand curve for a bread-exporting country.

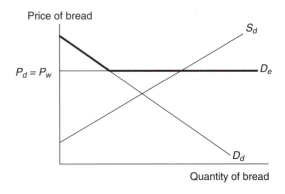

FIGURE 2.12
Effective supply curve for a wine-importing country.

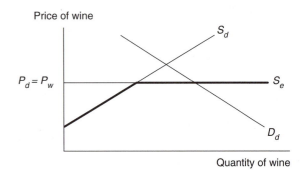

and the international supply curve, is perfectly elastic at the world price (P_w). The domestic price, P_d, is determined by the intersection of the effective supply curve, S_e, and the demand curve.

One last important topic is the explanation of the amount that a country will import or export at a given price. These curves are called *excess demand* and *excess supply curves*, and they are derived directly from the domestic supply and demand curves. Excess demand is obtained by taking the difference between what domestic consumers purchase and what domestic producers supply at a given price. Thus, excess demand, *ED,* is the horizontal subtraction of the domestic supply curve from the domestic demand curve. This is shown in Figure 2.13.

At the autarkic price, excess demand for the good will be zero, since the quantity of domestic supply is exactly equal to the quantity of domestic demand. At lower prices, though, excess demand is positive because domestic consumers will consume more than domestic producers will supply. This gap will be made up by imports (for example, the length of *BC* in the figure is exactly $Q_C - Q_P$). Thus, the excess demand curve slopes downward from the autarkic price with a slope that is flatter (more elastic) than the domestic demand curve. The excess demand curve is more elastic because it captures two effects as domestic price falls: (1) less is consumed, and (2) more is produced. The following box shows the relationship between domestic supply and demand elasticities and excess supply and demand elasticities.

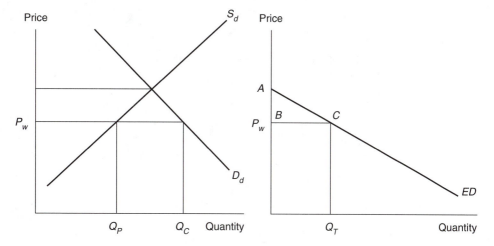

FIGURE 2.13
Graphical derivation of the excess demand curve.

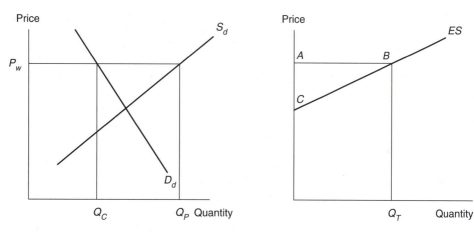

FIGURE 2.14
Graphical derivation of an excess supply curve.

Similarly, Figure 2.14 shows the derivation of an excess supply curve. Again, at the autarkic price, excess supply is zero because the quantity of domestic supply equals the quantity of domestic demand. At higher prices, excess supply, *ES*, is positive because domestic producers will supply more than domestic consumers will consume. This gap will be made up by exports (for example, the length of *AB* in the figure is exactly $Q_P - Q_C$). The excess supply curve will have a similar shape to the domestic supply curve, except it is flatter (more elastic). This is due to the two effects captured by excess supply as domestic price increases: (1) more is produced, and (2) less is consumed.

Derivation of the Excess Demand Curve

By definition the excess demand, M, is the difference between domestic demand, D, and supply, S.

$$M = D - S$$

Taking the derivative of each side with respect to price, P,

$$\frac{dM}{dP} = \frac{dD}{dP} - \frac{dS}{dP}$$

$$\frac{P}{M}\frac{dM}{dP} = \frac{dD}{dP}\frac{P}{M} - \frac{dS}{dP}\frac{P}{M}$$

$$\frac{P}{M}\frac{dM}{dP} = \frac{dD}{dP}\frac{P}{M}\frac{D}{D} - \frac{dS}{dP}\frac{P}{M}\frac{S}{S}$$

$$\epsilon_m = \epsilon_d \frac{D}{M} - \epsilon_s \frac{S}{M}$$

Where ϵ_m, ϵ_d, and ϵ_s are the excess demand, demand, and supply elasticities, respectively. Since ϵ_d is negative and ϵ_s is positive (and D, S, and M are positive), ϵ_m will be a larger negative number than ϵ_d. Import demand functions are always more elastic than ordinary demand functions. In a similar way, export supply functions are always more elastic than ordinary supply functions.

The excess supply and demand curves can also be used to show the gains from trade. If the world price in Figure 2.13 is P_w, the country in question will import Q_T from the rest of the world because there is an excess demand. The importing country's gain from trade can be measured as area ABC, which is exactly equal (by construction) to the gains from trade using the domestic supply and demand curves. Similarly, if the world price in Figure 2.14 is P_w, the country in question will export Q_T to the rest of the world because there is an excess supply. The exporting country's gain from trade is again measured by the area ABC, which is exactly equal to the gains from trade using the domestic supply and demand curves.

INTRODUCING THE WORLD MARKET AND LARGE COUNTRIES

The single good, small country analysis assumes that there is a well-functioning world market that determines world price, and the small country uses that world price to determine how much it should produce and consume. If there is

The Law of One Price

With perfect competition, no externalities, no trade barriers, and no transportation costs, the Law of One Price prevails. Because of arbitrage opportunities, traders will ensure that the good's price will be identical among all countries. If any difference prevails, pure economic profit could be obtained by purchasing the good in one country (with the low price) and shipping it to another country (with the high price), since all transportation and transaction costs are zero.

perfect competition throughout the world and no externalities, trade barriers, or transportation costs, this world price will be the true value of the good in question. The price will reflect the costs of producing the good and the value the good has in consumption. Under these assumptions, world market equilibrium occurs where the horizontal sum of all domestic supply curves intersects the horizontal sum of all domestic demand curves. It's logical that world price would be where world supply (or marginal costs) equals world demand (or marginal value).

Another way of looking at world market equilibrium is by using the excess supply and demand curves. Use of the excess curves will make it easier to envision the effects of relaxing assumptions concerning trade barriers and transportation costs. In this case, world market equilibrium occurs where the sum of the excess supply curves for exporters intersects the sum of the excess demand curves for importers. This equilibrium is shown in Figure 2.15, along with the surpluses (gains) that world trade generates. Total world welfare is maximized at a price of P_w and a quantity traded of Q_T. World importing countries gain the area ABC, while world exporting countries gain the area BCD.

A convenient simplification of the world market involves a situation where there is a two-country world (an exporting country and an importing country) and one good.[4] This situation is depicted in a three-panel diagram, as shown in Figure 2.16. The left panel depicts the domestic supply and demand situation in the exporting country, the right panel depicts the domestic supply and demand situation in the importing country, and the middle panel depicts the (excess) supply and (excess) demand situation in the world market. The excess supply

---

[4]The assumption of one importing and one exporting country is easily relaxed. However, the graphical analysis becomes much more cumbersome. One can simply view the export supply curve being the sum of many countries' export supply curves and the import demand curve being the sum of many countries' import demand curves. Results for individual countries would be obtained by going to the individual ES and ED curves by country. The results of various policies would be similar to the results outlined in this chapter.

FIGURE 2.15
World equilibrium using excess supply
and demand curves.

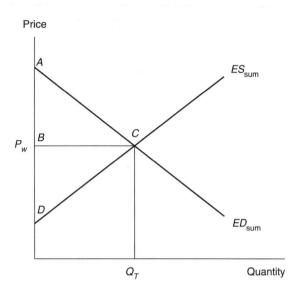

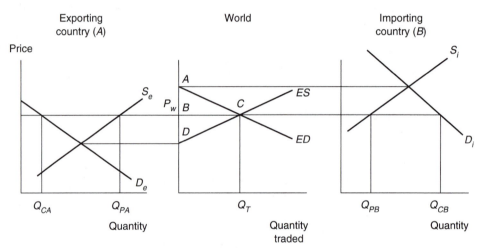

FIGURE 2.16
Trade equilibrium in the exporting country, importing country, and world.

function in the middle panel is derived from the left panel, and the excess demand function in the middle panel is derived from the right panel.

World market equilibrium is shown at P_w associated with exports and imports of Q_T. With no trade barriers and transportation costs, P_w will prevail in both countries and the excess supply in country A ($Q_{PA} - Q_{CA}$) will exactly equal excess demand in country B ($Q_{CB} - Q_{PB}$). World prices are equal to world opportunity costs, and world welfare from trade is maximized at area ACD in the middle panel (area ABC going to importers and area BCD going to exporters). There are no price differences between countries because transporta-

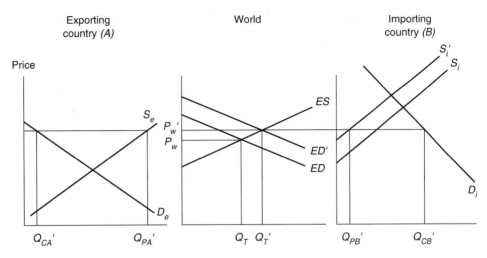

FIGURE 2.17
Effects of a drought in the importing country.

tion and transaction costs are zero, so arbitrage will ensure that there is no difference in price between countries.

It is very useful to investigate what happens to equilibrium when there is a shift in supply or demand curves. This is called *comparative static analysis,* which is easily performed using the three-panel framework. If the good in question is bread and there is a drought in the importing country (reducing wheat production), the supply curve in the right panel would shift to the left, to S_i', as depicted in Figure 2.17. The excess demand curve in the middle panel would shift upward to ED', forcing world price to increase (to P_w') and the quantity traded to increase (to Q_T').

The effects on the two countries could be traced back by using the new world price. Quantity supplied and demanded in the exporting country would move to Q_{PA}' and Q_{CA}', respectively; and quantity supplied and demanded in the importing country would move to Q_{PB}' and Q_{CB}', respectively. By construction $Q_{PA}' - Q_{CA}' = Q_{CB}' - Q_{PB}'$.

Notice that in this comparative static analysis, only one curve shifts. This is typical of comparative static analysis, and it highlights the important distinction between shifting curves and moving along curves. In the above example, the drought only shifted the supply curve and the excess demand curve in the importing country, but it caused a movement along the other curves. One must be careful in comparative statics to only shift the appropriate curves, otherwise the analysis will become confused and the results erroneous. Table 2.2 summarizes the effects of increased supply and demand by each country on world price and quantity traded. Decreases in supply and demand will have the opposite effect on world price and quantity traded. Interested readers can verify these impacts using a three-panel analysis.

TABLE 2.2	Effects of Increased Supply and Demand on World Price and Quantity Traded		
Shift		Effect on P_w	Effect on Q_T
Increase in importing country supply		P_w down	Q_T down
Decrease in importing country supply		P_w up	Q_T up
Increase in exporting country supply		P_w down	Q_T up
Decrease in exporting country supply		P_w up	Q_T down
Increase in importing country demand		P_w up	Q_T up
Decrease in importing country demand		P_w down	Q_T down
Increase in exporting country demand		P_w up	Q_T down
Decrease in exporting country demand		P_w down	Q_T up

INTRODUCING TRANSPORT COSTS

When the assumption of zero transport costs is relaxed, there is no such thing as the world price, which is true in the real world: for instance, there is no such thing as the world price of wheat. Instead, there are as many prices as there are locations in the world, reflecting transport costs from producing regions to consuming regions. In the simple two-country, one-good world being analyzed, there are two prices—the exporter's price and the importer's price—and a constant per unit transport cost from the exporting country to the importing country.

Because it is assumed that there are no other transaction costs (including trade barriers), arbitrage will guarantee that the price difference between the exporting country, P_E, and the importing country, P_I, will be no greater than transport costs, t:

$$P_I - P_E \leq t \tag{2.5}$$

Otherwise, traders would make a positive profit by buying wheat in the exporting country and selling wheat in the importing country.

If the price difference between P_E and P_I is less than transport costs, then traders will not export wheat because their export revenue cannot cover transport costs. Therefore, if trade takes place, the importing country price must equal the exporting country price plus transport costs:

$$P_I = P_E + t \tag{2.6}$$

The addition of transport costs can be incorporated into the three-panel analysis a number of ways. It may be easiest to view transport costs as an added cost of exporting, which can be depicted as a downward shift of the excess demand curve in the middle panel (to ED'). This is shown in Figure 2.18 with the corresponding price in the exporting country, P_E, price in the importing country, P_I, new world trade, Q_T', and transport cost, t. In this case the gains from trade are reduced to the small triangles ABC (for the importing country) and DEF (for the exporting country).

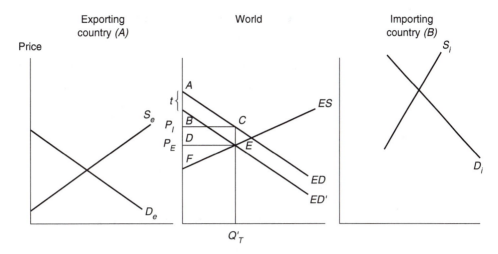

FIGURE 2.18
Introducing transport costs in a three-panel diagram.

The area $BCED$ is the transport costs for the good, which reflect the fact that world trade consumes transportation resources and therefore prices should reflect this.

Note that the exporting country has a lower internal price because of transport costs, while the importing country has a higher internal price because of transport costs. The incidence of the transport costs will depend on the shape of the excess demand and supply curves. If the excess demand curve is more elastic than the excess supply curve, the importing country price will increase less than the exporting country price will fall. The opposite is true if the excess supply curve is more elastic.

SUMMARY

1. World relative prices give the signals concerning what countries should produce and consume. In the Ricardian case, countries will specialize in the good where the country has a comparative advantage relative to other goods.
2. Countries gain from trade because they are able to concentrate on producing goods where they are relatively efficient and trade for goods where they are not efficient producers.
3. With concave production possibilities curves there will normally be incomplete specialization by countries: they will produce some quantity of each good. Gains from trade are measured through increased consumption that trade allows in one of the goods.
4. If a country exports a good, it is because the world price is above the country's autarkic price. Through exports, the country gains overall because its producers gain more surplus than consumers lose. If a country imports a good, it is because the world price is below the country's autarkic price. Through imports, the country gains overall because its consumers gain more surplus than producers lose.

5. Excess supply and demand curves are excellent ways to depict trade situations if countries are large. Comparative static analysis allows world price changes, trade flow changes, and gains from trade to be shown.
6. Relaxing the assumption of zero transportation costs results in less trade. Further, there are at least two prices that prevail in the world: price in the importing country (which is higher) and price in the exporting country (which is lower).

QUESTIONS

1. How would you extend the analysis of gains from trade to more than two countries? to more than two goods?
2. What evidence can you find that gains from trade exist?
3. Why are agricultural economists more interested in export and import volumes, rather than returns to production factors?
4. What wheat-exporting countries would you consider "large"? What wheat-importing countries would you consider "large"?

Chapter 3

Trade Policies of Importing Countries

Despite the clear gains from free trade, the world is characterized by trade barriers that make prices in importing countries higher than world prices. These trade barriers increase the price of imported goods and therefore change relative prices in the economy. As relative prices shift, production and consumption adjust and welfare of various groups changes.

This chapter covers the production, consumption, and welfare changes from various policies imposed by importing countries. The effects of trade policies on consumer and producer surplus are highlighted because they show the losses that come from these barriers. It is important to understand the trade-offs that occur when a country decides to pursue its pricing policies, rather than relying on world prices for its signal to producers and consumers.

MAJOR TYPES OF IMPORT BARRIERS

Import barriers can come in many forms, and countries have been increasingly creative in establishing new barriers over time. The earliest and most visible trade barrier is the simple import tariff (or tax). The *import tariff* can be a fixed amount per unit (a specific tariff) or a fixed percentage of the imported good's price (an ad valorem tariff). Generally, exporters can supply any amount to the importing country, but all imports are subject to the duty when they cross the border into the country.

Another common import barrier is the *import quota*, which restricts the quantity of a product that can enter the country. Sometimes these quotas are fixed on a country basis (e.g., the Philippines can ship only X tons of sugar to the United States). A firm must have an import license or visa before it is allowed to ship the product into the importing country. As will be seen later in this chapter, if the domestic price in the importing country is higher than the world price, then the import license has a value.

Import licenses can be auctioned to the highest bidder, sold for a fixed price, given to importing firms, or distributed through government agencies. The way these licenses are distributed can be very important in determining welfare effects.

The *tariff-rate quota* is a combination of the import quota and import tariff. In this policy, a fixed quantity or value of imports is allowed at a preferential tariff (sometimes zero) and all imports over that quota are subject to a higher duty. Often the higher duty is prohibitive such that imports above the quota are zero. This policy gives exporting countries access to the importing country's market, but the domestic price in the importing country can still be above the world price plus the preferential tariff.

In the last few decades, some governments have chosen to follow policies that keep some agricultural prices fixed in their country. There are two basic import policies that can allow this domestic policy to prevail: the government can control importing through *state trading* (so they import at the world price and sell at the fixed domestic price) or a *variable levy* system can prevail (where the import tariff is the difference between the fixed domestic price and the world price).

These are the major import policies that are used in the world today, and they are analyzed in this chapter. Technical barriers to trade are covered in another chapter because they are not easily analyzed in economic terms.

REASONS FOR TRADE BARRIERS

There are many reasons why a government might want to impose barriers on imports, but in most instances, these policies reduce the overall welfare of consumers and producers. In general, the government may want to redistribute welfare among the three major groups: producers, consumers, and the government. Governments often rely on import tariffs and other income from trade barriers to provide a substantial portion of the *government budget.* Collecting a duty on imported goods is an easier way to provide revenue than establishing and policing an income tax, and it is less visible than collecting a sales tax at retail outlets.

Some countries feel that certain products must be protected from international competition for national security reasons, and food is one of the products that is sometimes mentioned as important for national security. Governments often feel that *self-sufficiency* is important and that domestic producers should not be driven out of business by foreign competition. A similar argument for trade barriers is to *protect newly established businesses* during a critical phase in their development (the infant-industry argument). It can be shown, though, that in both of these cases, it is more efficient for the government to subsidize producers through direct payments than to distort prices through trade barriers.

Another important reason for import barriers is that businesses and farmers feel they must be *protected from import competition,* and the government acquiesces because of their political power. In essence, the government gets more benefit from an additional unit of welfare to producers than to consumers. Lobbying groups and the theory of collective action help explain this result. It is in the interest of producers and farmers to lobby hard for import protection because of the tremendous difference it can make in their income. In contrast, consumers are a much more diverse group that is not well organized. Consumers have some interest in freer trade, but most trade policies affect them less, per person, than producers.

The only legitimate economic reason for import barriers is if the *importing country is large* enough that it can use its market power to extract welfare from the rest of the world. In this case, the importing country is so large that if it restricts its imports, the world price will fall. This fall in world price can benefit the large importing country and increase its welfare. The importing country is extracting welfare from exporting countries by changing its importing pattern.

EFFECTS OF SPECIFIC IMPORT POLICIES

The analysis of each import policy will start from the free trade case and measure the losses resulting from various trade barriers. The welfare analysis will weight consumers, producers, and the government equally[1]. Unless otherwise noted, the small country case, zero transport costs, and perfect competition are assumed. Free trade is used as the baseline because free trade is generally the policy that maximizes the country's welfare. In each case the effective supply and demand curves are important to the analysis. The effective supply curve includes international and domestic suppliers, and the effective demand curve includes international and domestic demanders.

Import Tariff

The simplest and clearest import barrier is a tariff or tax on imports. The tariff can be a constant amount per unit of the product (specific tariff) or a constant percentage of the product's value (ad valorem tariff). Either way, under perfect competition

---

[1]With this assumption, $1 going to producers is valued the same as $1 going to consumers or to the government's treasury. Often decision makers value a dollar for producers more than a dollar for consumers.

FIGURE 3.1
The importing country with
free trade.

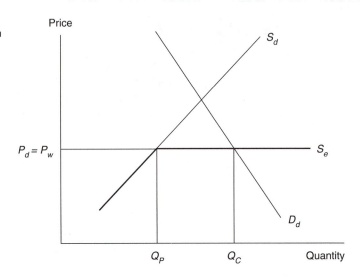

there will be a wedge between the exporter's price and the importer's price, even when transport costs are zero. The effects, however, are very similar to the impact of positive transport costs covered in Chapter 2. The distinction is that there are no resources that are consumed (transportation) to justify the difference between the import price and the export price. With an import tariff, the wedge is artificial.

If the country is small, the normal supply–demand picture will tell the entire story, since the country does not markedly influence world supply or demand. Under free trade, the country faces a perfectly elastic supply curve from the rest of the world (at P_w in Figure 3.1). When this function is added to the domestic supply curve, the effective supply curve, S_e, is the darkened line that includes the domestic supply curve below P_w and is horizontal at P_w. In equilibrium (assuming that the autarkic price is above P_w), the country produces at Q_P, consumes at Q_C, and imports $Q_C - Q_P$. The importing country price is P_d, which is identical to the world price.

If the country imposes a specific tariff of t per unit, the domestic price will move up to $P_w + t$ as long as the autarkic price is above $P_w + t$. Under perfect competition, importers are not making positive economic profits, so they must pass on the tariff fully to the importing country. This new internal price changes production and consumption patterns in the importing country, as shown in Figure 3.2, but it does not change the world price because the importing country is small (the added production and lower consumption have no significant influence on world markets). The effective supply curve is now the domestic supply curve below $P_w + t$, and it is horizontal at $P_w + t$. Producers react to the new price and increase production to Q_P', consumers react by decreasing consumption to Q_C', and imports fall to $Q_C' - Q_P'$.

Figure 3.2 can be used to see the gains to producers, consumers, and the government. Producers benefit from the tariff because they receive a higher price for their output, and this higher price also encourages them to produce more. Producer surplus increases by the area $ABED$. The government receives t for every

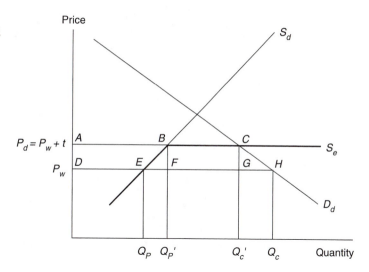

FIGURE 3.2
Effects of an import tariff.

unit imported, so it has revenue of *BCGF*. Since this revenue can be used to fund government services, it is also a benefit to the economy. Consumers are the losers because they suffer an increase in price for the good, which also encourages them to consume less. The loss in consumer surplus is *ACHD*, an area that exceeds the other two areas combined.

The country's welfare falls by the two triangles *BFE* and *CHG*. The first triangle, *BFE*, is often called the *production loss* because it measures inefficiencies caused by increased production by domestic producers. The tariff has encouraged production of the good by domestic producers, whose marginal cost is above that of the more efficient international producers. World marginal cost is at P_W, but each unit produced beyond Q_P in the importing country has a marginal cost above P_W. When these extra costs are summed over the increased production, the triangle *BFE* is obtained.

The second triangle, *CHG*, is often called the *consumption loss* because it measures inefficiencies caused by reduced consumption. The tariff sends the wrong signals to consumers because the new price does not reflect the world's opportunity cost for the good. Consumers are being sent a false signal that the good is more scarce, and this signal causes them to reduce consumption. Consumers are losing surplus because they would have consumed more than Q_C' and received added welfare if the price signals were accurately reflecting world opportunity costs. When this added value is summed over consumers, the triangle *CHG* is obtained.

As stated earlier, many countries are considering moving from restricted trade to freer trade. When eliminating a tariff, a scheme could be invented that would pay domestic producers for their losses from freer trade and there would be enough gain on the part of producers that the welfare of the economy would still improve. Hayami proposed a scheme for liberalizing Japan's beef import system in 1979. It would provide a deficiency payment for beef producers, while allowing consumers access to imported beef at a much lower tariff. Japan did finally liberalize

Japanese Beef Trade Liberalization

The Japanese beef market was controlled by the Livestock Industry Promotion Corporation (LIPC), a quasi-government organization, until July 1991, when private traders were allowed to handle all imported beef directly. The termination of the LIPC as a state trading entity was part of the Japanese Beef Access Agreement of 1988. Prior to then, the LIPC dictated the quality and origin of most Japanese beef imports, and licensed traders would then import the specific beef cuts that the LIPC requested. Only a small portion of the trade was allocated to specific traders, who could make exact orders on what beef they wanted.

The LIPC had specific quotas that it had negotiated with Australia and the United States (the two leading beef exporters to Japan), but the final allocation between these two countries was determined by administrative decisions, not the market. This was frustrating to beef traders because the Japanese beef market is far and away the most differentiated in the world. One can go to a Japanese steakhouse and order four to six different qualities (with four to six widely different prices) of the same beef steak.

The Japanese *wagyu* beef is the highest quality and most expensive. U.S. beef is typically within the middle portion of the quality spectrum (slightly below beef from Japanese dairy steers), while Australian and New Zealand beef (which is grass-fed) is normally viewed as the lowest quality (MATRIC). In 1977, the Australians controlled 85 percent of the Japanese beef import market (the United States had 9 percent), but by 1985, the Australians were down to a 62 percent share versus the United States' 30 percent share (Hahn et al.). Some economists felt that the Japanese were allowing the U.S. share to increase for political reasons (Alston et al.).

its beef imports in 1986 (a very controversial move on the government's part), but there was no deficiency payment offered (Coyle). Nonetheless, Hayami's analysis showed that there were huge benefits available for Japanese consumers if the objections of the small number of beef producers could be overcome.

Import Quota

The effects of the import tariff shown in Figure 3.2 can be identical with the effects of an import quota at a given point in time (when there are no shifts in supply and demand). The producer and consumer surplus changes with the tariff are the same as though the government announced an import quota of $Q_C' - Q_P'$ units. This import quota would produce a scarcity in the domestic market that would force the importing country's price to $P_W + t$, and producers would gain and consumers would lose the surplus amounts. If the government sold the import licenses (the right to import) to traders in a perfectly competitive market, the government

Before liberalization, all Japanese beef imports were subject to a 25 percent tariff, but the price of Australian beef on Japanese wholesale markets was normally much higher than a 25 percent tariff implied. These implied tariffs on imports were needed because the LIPC was supposed to keep domestic beef prices in Japan within a specific range. Some of these implied tariffs, or "rents," were captured by the LIPC's resale price (the government sold imported beef for much more than its purchase price), but the rest accrued to Japanese butchers, who simply made a high margin on any Australian beef they were allowed to buy through the LIPC (Hayami). Profits for the LIPC were used to help domestic beef producers through subsidized credit, deficiency payments for calves, and other programs.

In July 1991, the LIPC became simply a research and reporting agency of the Japanese government, turning over importing to private traders. Beef imports into Japan then became subject to a 70 percent tariff for the period July 1991 through June 1992, a 60 percent tariff for the period July 1992 through June 1993, and a 50 percent tariff after June 1993. During the Uruguay Round of the GATT negotiations, the Japanese pledged to lower their beef tariff to 38.5 percent.

Eliminating the LIPC, and the concomitant reductions in tariffs, has had a tremendously positive impact on U.S. beef exports to Japan. In 1987, U.S. beef exports totaled $562 million—or 397 million pounds on a carcass weight basis (Hahn et al.). In 1997, U.S. beef exports totaled $2.31 billion (or over 1 billion pounds on a carcass weight basis). It is amazing what private traders can do to increase markets if they are allowed.

would capture the revenue as in the import tariff case (area *BCGF* in Figure 3.2), because the quota licenses would be sold at t per unit.

The distribution of import licenses is not always accomplished through a bidding process in a perfectly competitive environment. Often the government will distribute licenses to traders or even producers within an exporting country based on past history. The fees for the licenses (if they exist at all) could be fixed and much less than what the licenses are worth (which is t per unit in this example). The country loses any surplus that goes to international traders if the licenses are priced below t per unit, so there is potential for an additional welfare loss under an import quota. Further, the country may not value highly any surplus captured by large importing firms (even though they are domestically based) by buying at the low world price and selling at the high domestic price.

The most important difference between an import tariff and an import quota involves the effective supply curves. As stated earlier, the effective supply curve under the import tariff case is perfectly elastic at $P_w + t$. This is not true under the import quota case. Figure 3.3 shows the effective supply curve under the import

FIGURE 3.3
Effective supply under an import quota.

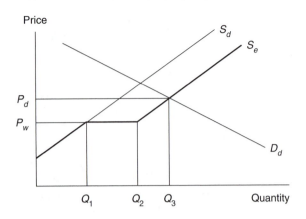

quota case—the darkened line in the figure. The effective supply curve is the domestic supply curve at prices below P_w, perfectly elastic from that point until the import quota is exhausted, and it has the slope of the original domestic supply curve above P_w.

At prices below the world price, the effective supply will come from domestic producers only. At P_w, the world is willing to supply any amount that the country will allow, but the country will allow only $Q_2 - Q_1$ (the amount of the import quota). The only way that the domestic market can be supplied product after the import quota has been exhausted is for domestic producers to increase production, which requires a higher domestic price. That is why the S_e curve has a positive slope at quantities above Q_2. In Figure 3.3, domestic production is $Q_1 + Q_3 - Q_2$, domestic consumption is Q_3, and imports are $Q_2 - Q_1$. The domestic price is P_d, and the world price is P_w.

Obviously because the effective supply curves are different between the import tariff and quota, the effects of shifts in domestic supply and demand will differ. Figure 3.4 highlights these differences where the import tariff case is shown in the left panel and the import quota case is shown in the right panel. Both panels have initial consumption at Q_3, and initial domestic production is identical ($Q_p = Q_1 + Q_3 - Q_2$). If the domestic demand curve (D_d) shifts outward in the import tariff case to D_d', the level of imports rises by the amount that the demand curve shifts. In the import quota case (assuming that the quota is binding), shifting the domestic demand curve to D_d' increases the domestic price to where the effective supply curve intersects the demand curve; imports will not change.

If the domestic demand curve shifts to the left in the import quota case, it is possible that the market will be in the region where P_w prevails but there are still imports (between Q_1 and Q_2). In this case, the import quota is not binding and import licenses have no value because the domestic price is identical with the world price. So there is a region of consumption levels that would generate import levels consistent with free trade, even though there is an import quota policy. This possibility does not exist with the import tariff case: all imports face a tariff.

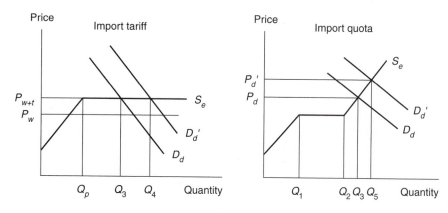

FIGURE 3.4
Differential effects of a demand shift with import tariffs and quotas.

Shifts in the domestic supply function will affect only the lower part of the effective supply curve under an import tariff, but a domestic supply shift will affect the entire effective supply curve under the import quota case. Under an import tariff case, as long as there are imports, shifting the domestic supply curve will change only the ratio of domestic production versus imports. Domestic supply shifts in the import quota case may change import levels (if the quota becomes nonbinding) and will definitely change domestic price.

World Price Stability under Tariffs and Quotas

Import tariffs and quotas also have different effects on world price stability. When the world price changes, it is a signal that the relative scarcity of the product in question has changed. If the world price increases, the product has become more scarce and that sends signals for producers to produce more and consumers to consume less. Under an import tariff, the importing country allows those world price signals to be transmitted (through a higher internal price) to their domestic producers and consumers. Because the price is allowed to change in the importing country, production and consumption will adjust.

Under an import quota, those signals are not allowed to be transmitted to domestic producers and consumers, as long as the quota is binding. If the world price is lower than the domestic price in the importing country, the importing country will not change its production and consumption decisions because the domestic price doesn't change. The importing country has shielded itself from outside disturbances. Remember, however, that the domestic price in the importing country changes if the domestic supply and demand conditions change. If the domestic situation is more volatile than the world situation, the importing country will experience more domestic price fluctuations under an import quota than under an import tariff.

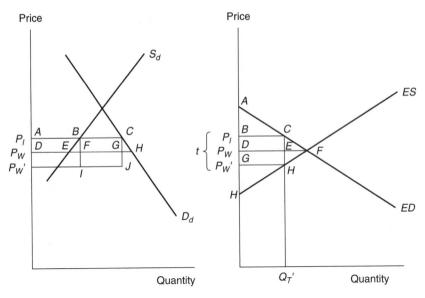

FIGURE 3.5
Import tariff for a large country.

The Large Country Case and Optimal Tariff

If the small country assumption is relaxed, the importing country's policies can affect the world price. Any policy that reduces the country's importation will reduce world price, and an import tariff certainly falls into this category. If the importing country is large, one must use at least two panels from the three-panel diagram to see the full impact of an import tariff.

Figure 3.5 shows that when the importing country institutes a specific tariff of t per unit, the difference between the world price and the import price is t units. The world price falls from P_w to P_w', and the price in the importing country increases to P_I. Notice that the importing country's price increases by less than the amount of the tariff. In essence, the importing country forces the rest of the world to pay part of the tariff through a lower world price. The quantity imported falls to Q_T'.

The importing country's welfare changes associated with the large country case are quite interesting because the rest of the world pays part of the tariff. Surplus measures for the importing country are indicated in the left panel. The tariff causes producer surplus to increase by area $ABED$ in the left panel, consumer surplus to decrease by area $ACHD$, and government revenue of area $BCJI$. The production and consumption losses, areas BFE and CHG, respectively, are smaller than under the small country case because the new world price is lower. It is unclear whether the large country is worse off from the tariff. One must compare the sum of areas BFE and CHG, the losses, with the area $FGJI$, the area gained because world price fell. If the latter is greater than the former, then the large country has gained from imposing the import tariff.

FIGURE 3.6
Optimal import tariff.

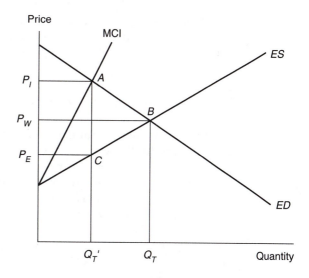

The right panel shows gains/losses for the importing and exporting countries. The panel clearly indicates that the exporting country has lost because of the import tariff. Specifically, the exporting country has lost area *DFHG*. The importing country's gain from trade under the import tariff is area *ACHG*, representing, relative to free trade, a gain of area *DEHG* and a loss of area *CEF*. If the former area is greater than the latter, then the importing country has gained from the import tariff.

As long as the importing country faces an upward-sloping *ES* curve, there is an import tariff that will increase welfare for the importing country. The import tariff that maximizes the large country's gain from trade is called the *optimum tariff*. An optimum tariff exists for any large country because its marginal cost of import (MCI) is always greater than the import price. The reasoning behind this phenomenon is similar to the reasoning behind the marginal revenue curve lying below the demand curve in the monopoly case. The only way a large importing country can get additional output from the rest of the world is to bid up the price. Bidding up the price encourages other countries to produce more and consume less. However, the increased price accrues not only to the last units traded internationally but also to all other units traded. Thus, it costs the large importing country more than the world price to increase imports because it costs the country more than the world price to buy additional units.

The marginal cost of import (MCI) curve is shown in Figure 3.6, lying above the *ES* curve. If the importing country took advantage of its size in the world market, it would import until the *MCI* curve intersects the *ED* curve (where the country's marginal costs equal marginal benefit). The quantity imported would be Q_T' (less than Q_T), the price for exporters would be P_E (lower than P_W), and the importing country's price would be P_I. The large importing country's gain from the tariff (which was shown in Figure 3.5) will be positive, while the exporters (and the rest of the world) would lose. Overall, the world would lose the area *ABC* because P_E and P_I are different than P_W, and P_W reflects the true scarcity of the good.

Note that the optimal tariff for a small importing country is zero because it faces an *ES* curve that is perfectly elastic. Any tariff imposed by a small country will result in lower welfare for the country.

Fixed Internal Prices by Importing Countries

The real reason behind trade barriers for many importing countries is to support domestic producer incomes. Governments weight the producer's welfare more highly than they weight consumers' welfare, especially for agricultural products. Often governments decide what domestic policies they wish to use to support domestic producers and then determine a trade policy that will allow them to attain their domestic goals. Often these domestic policies involve guaranteed or minimum producer prices. If an importing country wants a minimum producer price that is above the world price, they must have some trade barrier to ensure that imports cannot be sold below the minimum producer price. A variable levy will perform this function.

A *variable levy* is a tax applied to imports to ensure that they cannot enter a country below a fixed minimum level (often called a *threshold price*). Mathematically, the variable levy is simply the difference between the minimum import price (P_T) and the world price.

$$L = P_T - P_w \tag{3.1}$$

The levy is variable because the world price can vary on a daily basis. This variable levy system guarantees that the higher domestic price will not be undermined by imports. The European Union uses a variable levy system for its major agricultural products (e.g., corn, wheat, pork, beef). Every day importers submit bids on the price of their product and the customs authority subtracts the lowest bid from the threshold price (established for that year) to obtain the variable levy for that day. All imports of that product are assessed the variable levy for that day, ensuring that imports are at least as expensive as domestic products.

A variable levy system makes supply and demand points below the threshold price irrelevant because those prices cannot prevail. In other words, the effective supply and demand curves, S_e and D_e, become vertical (or perfectly inelastic) below the threshold price, as shown in the left panel of Figure 3.7. This makes the excess demand function for the importing country vertical below the threshold price too, as shown in the right panel. The gains for producers and the government, and losses to consumers, can be shown easily in a manner similar to the import tariff case (Figure 3.2).

The main difference between the import tariff case and the variable levy case is highlighted when world supply and demand conditions change. If the world price changes in the import tariff case, there will be a change in the importing country's price, but that is not true in the variable levy case as long as the world price stays below the threshold price. In this regard, the variable levy is more similar to an import quota because world prices are not transmitted to the importing country.

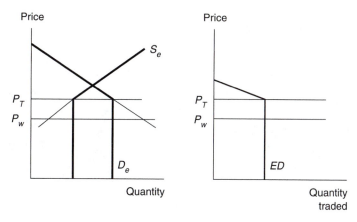

FIGURE 3.7
Supply and demand curves with a fixed internal price.

Tariff-Rate Quotas

The tariff-rate quota (TRQ) has become more popular in recent years as countries have attempted to lower their trade barriers over time. A TRQ allows a certain amount of imports at a lower tariff (sometimes zero), with imports above the quota assessed at a higher tariff. This concept doesn't make much sense until one considers the context of highly restricted trade. The TRQ allows some minimum access for imports, but domestic prices can still be very high because of the higher tariff for imports above the quota.

Often countries are in a situation where they guarantee that they will simplify their import policies and lower trade barriers significantly over time. The way they simplify them is to move from import bans, licenses, quotas, and other restrictive regulations that may not be completely clear and straightforward to a simple import tariff. In that way, prospective importers know that they will face an import tax, but they will not need to work with a government bureaucracy to obtain a license or go through myriad procedures to receive approval. Importers know that they will be able to get their product through customs as long as they pay the tax. This process of moving from many import barriers to a simple tariff is called *tariffication*—changing all import barriers into their tariff equivalent.

After the country has moved to a simple tariff, it is easy to measure how fast trade barriers are being lowered because one simply measures how the tariff changes over time. The tariff-rate quota is used to guarantee that there is increased access to the importing country's market through the liberalization process. The import quota might be set at some average import level over the previous few years and the lower tariff might be zero, guaranteeing that imports will be at least as large as they have been in previous years. Imports beyond the quota would be taxed at the new "tariffied" rate. This gives the exporting countries the benefit of not losing their past market when tariffication takes place, and gives them access to a larger market if their price can be competitive with the new tariff.

Tariff-rate quotas were very important in the early years of the North American Free Trade Agreement (NAFTA). Both the United States and Mexico agreed on TRQs for some products where the quota was based on the average of the previous three years' imports and the tariff was zero. Imports beyond the quota were taxed at a very high rate that reflected the tariffication of import policies at that time. Liberalization occurred through a progressive increase in the duty-free import quota and a progressive reduction in the tariff on imports beyond the quota. TRQs were also an important part of the 1994 GATT agreement, as will be discussed in Chapter 5.

State Trading

The final form of trade barrier covered in this chapter is when the government has a state-owned organization that controls importation. Normally, with state trading, the government organization will be the sole importer and will control the resale price of the imported product. This situation was rather common in the 1960s and 1970s, particularly in less developed countries, but the wave of privatization that has occurred during the 1990s has made state trading less common throughout the world.

When the government controls importation, many of the other policies covered in this chapter could be implicit in the organization's actions, making for a very restrictive trade policy. The state organization can control exactly how much is imported, since it is the sole importer. The actions of the government agency can serve as an import quota that can change through the agency's own simple actions or dictates. Further, since the government agency sells the imported products in the local market (either directly to consumers through its own stores or to wholesalers), the agency can impose implied tariffs (or subsidies) that will affect the local price of the product.

SUMMARY

1. Trade barriers are used for many purposes, but their intent is normally to protect domestic producers and therefore shift welfare from consumers to producers.

2. Import tariffs increase the domestic price for a good, encouraging production and discouraging consumption. Producers gain, consumers lose, and the government gets a revenue source. However, with a small country, consumer losses are greater than producer and government gains.

3. An import quota specifies the maximum amount that can be imported into a country. Its effect is similar to the import tariff (domestic prices increase if the quota is binding), but the dynamics are different. If there are domestic supply or demand shifts when the quota is binding, the domestic price will also change, whereas it will not change with an import tariff.

4. A large country can gain from an import tariff by significantly reducing world trade. In this case, the large country extracts welfare from other countries by lowering world price, which will reduce welfare of exporting countries. The large country's gain in producer surplus and government revenue exceeds the losses incurred by its consumers.

5. A variable levy is a tax imposed on imports that makes up the difference between a fixed domestic price and the world price. The welfare effects of a variable levy are the same as an import tariff at a point in time, but if domestic supply and demand curves change, there will be no change in the country's importing pattern.

6. A tariff-rate quota is a means to guarantee that some minimum amount (the quota) will be imported at a lower tariff, but imports above the minimum are subject to a higher import tariff. This policy has been popular in recent trade negotiations (GATT and NAFTA).

QUESTIONS

1. Why have import policies become more complex over time?
2. The United States imposes relatively high tariffs on sugar and dairy imports. Why do you think the United States follows such policies?
3. Why would a government weigh producer surplus more than consumer surplus? How do you think those weights have changed over time, and how will they change in the future?
4. Can you think of instances where countries have imposed import tariffs to take advantage of their market power? Explain.
5. What would the welfare and trade impacts be if the country was large and it imposed an import quota?

REFERENCES

Alston, Julian, Colin Carter, and Lovell Jarvis. "Japanese Beef Trade Liberalization: It May Not Benefit Americans." *Choices* (Fourth Quarter, 1989): 26–30.

Coyle, William T. "The 1984 U.S.–Japan Beef and Citrus Understanding: An Evaluation." Foreign Agricultural Economic Report No. 222. Economic Research Service. U.S. Department of Agriculture, July 1986.

Hahn, William, Terry Crawford, Linda Bailey, and Shayle Shagam. "The World Beef Market—Government Intervention and Multilateral Policy Reform." Staff Report No. AGES 9051. Washington, DC: U.S. Department of Agriculture, Economic Research Service, August 1990.

Hayami, Yurijo. "Trade Benefits to All: A Design of the Beef Import Liberalization in Japan." *American Journal of Agricultural Economics* 61(1979): 342–7.

Midwest Agribusiness Trade Research and Information Center (MATRIC). "Meat Marketing in Japan: A Guide for U.S. Meat Exporting Companies." Des Moines and Ames, IA: MATRIC, 1990.

Chapter 4

Trade Policies of Exporting Countries

Policies that drive a wedge between world prices and domestic prices are not unique to importing countries: exporting countries often use policies that affect their trade patterns. Again, these policies are normally used to meet domestic policy objectives, which vary markedly among exporting countries. The common thread, though, is that export policies are normally geared toward increasing exports, rather than keeping production at home.

This chapter covers the main export policies used by countries to influence trade. The assumptions for the analysis of each export policy covered in this chapter will be the same as for the import policies. The analysis begins with the free trade case and measures the losses in welfare by using equal weights for consumers, producers, and the government. Unless otherwise noted, the small country case, zero transport costs, and perfect competition are assumed.

EXPORT SUBSIDY

The first export policy analyzed, the export subsidy, is common for agricultural products in more developed countries such as the United States. An export subsidy allows the domestic price in the exporting country to exceed the world price. This policy is always welfare-reducing if the country weighs producer surplus, consumer surplus, and government revenue equally. Yet it is a common policy for raw agricultural products.

The export subsidy case for a small country is presented in Figure 4.1, where domestic supply and demand curves are drawn with a price in the exporting country, P_d, that is above the world price by t units, the per unit amount of the export subsidy. Presumably the government in the exporting country has determined that its domestic price must be t units above the free trade price, so the only way that

FIGURE 4.1
Domestic supply and demand with an
export subsidy (small country).

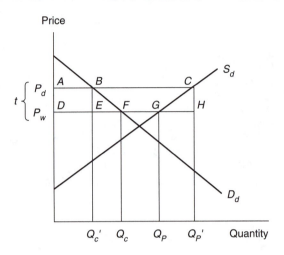

the exporting country can get rid of its surplus ($Q_P' - Q_C'$) at the prevailing domestic price is through an export subsidy of $P_d - P_W$. Note that in this example domestic consumers pay the nonsubsidized price of P_d. If the government chooses to allow domestic consumers to purchase at the world price (through a subsidy), then exports would fall by $Q_C - Q_C'$.

Domestic producers benefit from the export subsidy because their surplus increases by *ACGD* and consumers lose by *ABFD*. The government is also a loser in this case because *BCHE* must come out of the government budget to pay exporters to supply $Q_P' - Q_C'$ to the world market. The net losses to the exporting country are *BFE* and *CHG*. In the world market, the small country faces a perfectly elastic excess demand curve, so the world price does not change with an export subsidy and there are no significant benefits to the rest of the world.

The large country case is even more welfare-reducing for the exporting country because the export subsidy results in significantly more exports for the world, which reduces the world price. This result is shown in Figures 4.2 using two panels of the three-panel diagram. The export subsidy causes the excess supply curve to shift downward by the amount of the subsidy, lowering world price to P_W'. The domestic price in the exporting country goes up by less than the amount of the export subsidy because the world price has fallen (rather than $P_W + t$, the new price in the exporting country, P_d', is $P_W' + t$). Consumer losses (*ABFD*) and producer gains (*ACGD*) are much less than when the country was small. The government still loses the export subsidies that are paid (*BCJI*). The triangular losses for consumption (*BFE*) and production (*CHG*) are less, but the net loss to the country (*BFE + CHG + EHJI*) is greater. Clearly the net losses from the export subsidy are greater for the large country (Figure 4.2) than for the small country (Figure 4.1).

Despite this result, some large exporting countries, including the United States, continue to subsidize exports. One potential reason is that they value producer surplus more than consumer surplus or government revenue (cost).

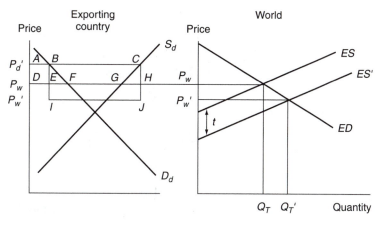

FIGURE 4.2
Export subsidy when the country is large.

If you weigh producer surplus high enough (relative to consumer surplus) in the country's social welfare function, the country can gain from the export subsidy. Over time, however, the United States has moved from general export subsidies to targeted export subsidies in order to combat subsidies by other countries, specifically the European Union. These targeted export subsidies reduce government costs because the subsidies are given only for certain markets where the United States competes with export subsidies from other countries (particularly the E.U.).

The European Union is forced to subsidize exports because of the domestic prices decided upon through the Common Agricultural Policy (CAP). Recall the variable levy system that prevails for many agricultural products in the E.U. (Chapter 3). When the CAP was formulated in 1958, the supply–demand situation was such that the autarkic price was above the threshold price (guaranteed minimum price), so the E.U. consumed more than it produced. Over time, because of higher internal prices, technological advances, efficiency gains, and other factors, the E.U. supply–demand situation changed such that the autarkic price was below the threshold price, so the E.U. produced more than it consumed. This forced them to subsidize exports.

The two situations that the E.U. has faced with its CAP over time are shown in Figure 4.3. The left panel is the situation when the E.U. is a net importer with a variable levy system (this is identical to the left panel in Figure 3.7) and the government receives revenue from the variable levy. The right panel is the situation when the E.U. moves to a net export position with a variable levy system. The supply curve has shifted out enough that its relevant portions lie to the right of the demand curve, making it necessary for the E.U. to export if the internal price is to remain at P_T. The E.U. is forced to provide a subsidy of $P_T - P_W$ if exports are to occur. Without exports, there is no way to get rid of the surpluses through market forces.

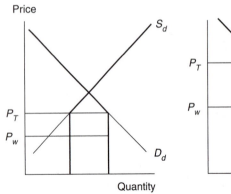

Country is importer

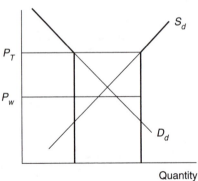

Country is exporter

FIGURE 4.3
Effective demand and supply curves with a variable levy.

EXPORT TAX

Export taxes can be collected directly from exporters or indirectly through a government marketing board that pays producers a price lower than the world price. Export taxes are more prevalent in less developed countries because they are an easy way for the government to obtain money from one of the few viable industries in the economy: agriculture. Governments may realize that this distorts economic decisions in the country, specifically distorting prices in agriculture, but these distortions are simply the cost of obtaining operating funds for the government.

The export tax forces the price in the exporting country below the world price by the amount of the tax. If a small country is involved, the domestic price will fall by the full amount of the tax, as shown in Figure 4.4 (domestic price falls from P_W to P_d), reducing exports from $Q_P - Q_C$ to $Q_P' - Q_C'$. Producers lose from this policy because they receive a lower price and they also cut back on production. Producer surplus falls by the area $AEHF$. Consumers are allowed to purchase the product domestically at the lower price, so they gain area $ABGF$, and the government collects revenue of area $CDHG$. The country as a whole loses the two familiar triangles, areas BCG and DEH, which are the consumption and production losses, respectively. There is no question that a small country will lose from an export tax. However, this is not true if the country is large. A large country can hold production off the market and force the world price higher, allowing the possibility that the country could gain if the production cutback is small relative to the world price rise.

The general case of an export tax with a large country is given in Figure 4.5. The exporting country's price (P_e) and the new world price (P_W') differ by the amount of the export tax, as shown in the right panel, reducing the quantity traded from Q_T to Q_T' and driving a wedge between the exporting country's price, P_e, and the new world price, P_W'. World price increases from P_W to P_W' because of the ex-

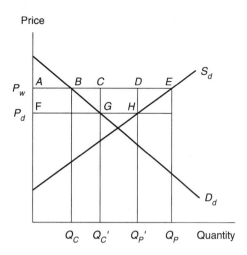

FIGURE 4.4
Effects of an export tax on a small country.

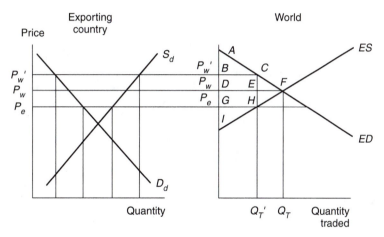

FIGURE 4.5
Effects of an export tax on a large country.

port tax. The figure does not show it, but the ES curve has shifted such that it moves through point C. The gains from trade for producers and consumers in the exporting country are area GHI, the gains for the importing country have shrunk to area ABC, but the exporting country's government has gained the area BCHG, making it possible for the exporting country to be a net gainer (in particular if area BCED is greater than the area EFH). There is no question that the importing country loses because of the export tax (area CEF + BCDE) and the world loses (area CFH).

The left panel, which shows the effects of the export tax on the domestic supply and demand situation, is similar to Figure 4.4 for the small country, except that the world price has increased, reducing the loss triangles and exposing an area of gain for the exporting country because of the higher world price. The gains from the policy can also be highlighted in the left panel, but that is left to the reader.

FIGURE 4.6
The optimum export tariff for a large country.

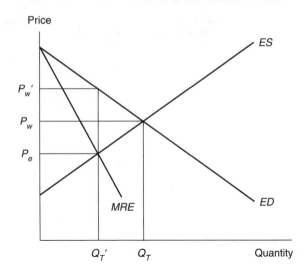

For every large exporting country there is an optimal export tax that will maximize the country's total welfare (producer surplus, consumer surplus, plus government revenue). This optimal export tariff exists because the marginal revenue from exporting for a large country is less than the world price. The only way the exporting country can increase exports is by reducing the price on all of its exports. Thus, the marginal revenue from exporting (*MRE*) curve lies below the *ED* curve faced by the large country, as shown in Figure 4.6. The welfare-maximizing exports for the large country are where *MRE* is equal to *ES* (or marginal cost of exporting). This level of exports is shown as Q_T', which is lower than Q_T, the free trade level of exports. The optimal export tax in Figure 4.6 is $P_W' - P_e$.

PRICE DISCRIMINATION

With an optimal export tax, the exporting country has extracted rents (or surplus) from the rest of the world by holding down exports and increasing the world price. The rest of the world is a net importer, so it is a net loser when the world price increases. This situation exists because the exporting country faces a downward-sloping excess demand curve. It can be shown that the more elastic (flatter) the *ED* curve, the smaller the optimal export tax because *MRE* lies closer to *ED*.

If the exporting country acts as a profit-maximizing (or surplus-maximizing) monopolist, the country will segment its markets and charge different prices to different markets. The exporting country can extract more rents from those countries that are willing to pay more because of demand and supply conditions. In this situation the exporting country would sell until *MRE* was equal to its marginal cost in every country. This decision-rule would result in lower prices in importing countries with a more elastic excess demand curve and higher prices in importing countries with a less elastic excess demand curve. The exporting country would be

Export Embargoes

The United States has used embargoes on agricultural exports periodically as a foreign policy tool. This chapter does not cover the economics of such a policy, but it obviously hurts agricultural producers. Two notable U.S. agricultural embargoes have occurred since 1970—the soybean embargo of 1974 and the Russian embargo of 1980. In both cases, the U.S. embargo merely forced importers to use other means to obtain their needed food supplies.

In June 1973, the U.S. price of soybean meal was more than three times that of the previous year. There was tremendous political pressure to control this serious price increase, so the United States placed an embargo on all oilseed exports on June 27. Five days later, the United States announced an export licensing scheme that lasted until October 1973. The U.S. Department of Agriculture's commissioned study found that there were little short-term or long-term effects of the embargo. Yet some economists argue that the Japanese began to invest directly in Brazil's soybean industry after that time (Tweeten).

The 1980 embargo on U.S. agricultural sales to the Soviet Union was imposed as punishment for its invasion of Afghanistan. The embargo lasted from January 1980 to April 1981. Grains were the most important commodities influenced by the embargo. The U.S. Department of Agriculture's commissioned study found that the embargo had little impact on Russian imports or consumption levels. The Russians simply changed their sources of agricultural supplies. The United States did implement agricultural programs to compensate U.S. farmers for lost Russian markets.

The conclusion from these embargoes, and from others as well, is that they rarely make much difference to the countries involved. Product distribution patterns might change during the embargo, but countries can usually find the products they need to feed their people. After the embargo, normal commercial channels are followed and there is little change in the way business is conducted. Some importers might complain of supply reliability, but in the final analysis, they are mostly interested in finding products at the lowest price possible.

charging a higher price (and extracting more rents) from importing countries that were not price responsive, and charging a lower price in importing countries where there was more competition.

An example of this situation with one exporting country and two importing countries is shown in Figure 4.7. The left panel shows the excess demand curve and corresponding *MRE* curve for country 1 (an importing country), and the middle panel shows the same curves for country 2 (another importing country). The right

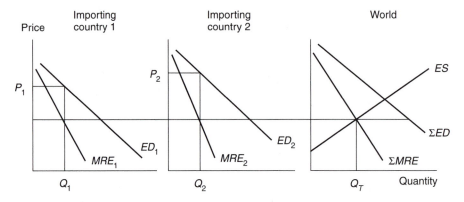

FIGURE 4.7
Price discrimination with two importing countries.

panel shows the horizontal summation of both curves for the two countries and the excess supply curve for the exporting country. If the exporting country exploits its market power and equates the summed MRE curve in the right panel with the ES curve (its marginal cost curve) to determine how much it supplies the two markets, the export levels would be Q_1 and Q_2 ($Q_T = Q_1 + Q_2$), with the corresponding prices at P_1 and P_2. Note that country 1 has a lower price because its ED curve is flatter (more elastic).

The key to the price discrimination case, though, is making sure that the product cannot flow among importing countries; otherwise, the price differences among countries can vary no more than transport costs. Easy markets for the exporting country to discriminate between, though, are the domestic versus the export market, since transport costs and import barriers often preclude sending the good back to the

Other Means of Export Enhancement

The U.S. Department of Agriculture has many other means at its disposal for increasing agricultural exports. These programs do not fit neatly under any of the categories covered elsewhere in this chapter. All of the programs here are authorized by the 1978 Trade Adjustment Act and are ways that the U.S. government subsidizes exporters or exporter organizations.

The oldest program is the Foreign Market Development Program (also known as the Cooperator Program), which provides matching support for producer organizations to promote exportation of products throughout the world. The federal funds for this program totaled $30 million in 1998. Through such organizations as the U.S. Wheat Associates, U.S. Meat Export Federation, and other groups, U.S. exports of agricultural products are promoted at trade fairs, exhibitions, media campaigns, and other events.

The General Sales Manager (GSM) 102 and 103 programs underwrite (guarantee) credit extended by U.S. commercial banks to less developed countries. Countries and commodities for GSM support are approved annually. The GSM 102 program extends credit for up to three years, while the GSM 103 program extends credit for up to ten years. This credit subsidy allows U.S. firms to sell products to countries that might not otherwise be able to afford imports. For fiscal year 1998, GSM 102 credits taken by U.S. exporters totaled $4 billion and GSM 103 credits totaled $56 million.

Finally, there is the Market Access Program (also known as the Targeted Export Assistance Program and the Market Promotion Program), which provides matching support to individual U.S. producers, exporters, trade associations, and others in their quest to develop international markets for their products. This program was originally designed to combat export subsidies by U.S. competitors in specific markets. However, it has been generalized and is available for U.S. products being shipped throughout the world, though countries and products are approved based on a competitive process. The Market Access Program had a $90 million budget for fiscal 1998.

exporting country. The domestic demand curve in the exporting country is also usually less elastic than the excess demand curves faced in international markets, so there is often fear that exporting countries will charge lower prices for exports than for domestic sales (the classic dumping case).[1] Most countries have laws against foreign companies dumping products on their markets (selling below the price they sell in their own country), but it is sometimes very difficult to prove.

[1]Figure 4.7 could be changed such that the middle panel depicts the domestic market and the left panel depicts the export market. Then the exporting country would have two prices, a high domestic price (P_2) and a lower export price (P_1).

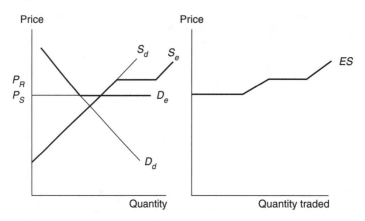

FIGURE 4.8
Price support system for an exporting country.

OTHER IMPORTANT POLICIES THAT AFFECT EXPORTS

There are two other policies followed by exporting countries that are not trade policies but have an effect on international trade. The United States has used both policies, but neither is currently followed for major commodities in the United States because of the 1996 Federal Agriculture Improvements and Reform (FAIR) Act. A discussion of these programs sheds additional light on how policies that are sometimes viewed as purely domestic can have large impacts on international trade.

Price Supports

The price support system was a major policy in U.S. agriculture from 1933 through 1985. The system had many facets and changed from year to year, but its basic concept was that a minimum price was established and the government purchased (or loaned farmers money to store) the product at that support price. Accumulated surpluses came out of storage at a release price. This program made sense as long as the price support was low relative to the long-run price of the product and accumulated surpluses didn't become burdensome.

The left panel of Figure 4.8 shows the effects of a price support policy on effective supply and demand curves. The effective demand curve (D_e, the darkened part of the demand line) is the domestic demand curve at higher prices, but it becomes perfectly elastic at the price support, P_S, because of government purchases. The effective supply curve (S_e, the darkened part of the supply line) is the supply curve below the release price, but it becomes perfectly elastic at the release price (P_R) until government carryover stocks are released, when it returns to the shape of the domestic supply curve. The excess supply curve in the right panel is consistent with the effective demand and supply curves. Carryover stock levels will determine the extent that the effective supply and the excess supply curves are flat at the release price.

The price support policy was common for grains, and the United States accumulated large stock levels in the 1950s, 1960s, and the late 1970s and early 1980s. The United States became known as a residual supplier of grains because importers knew that any shocks to grain prices would be absorbed by the United States. If world production was large for a particular year, the excess demand curve would fall to the price support level and the United States would purchase a high percentage of the U.S. crop, uplifting world prices. If world production was low for a particular year, importers could always go to the United States and purchase grain at the release price. One can imagine that U.S. exports fluctuated a great deal because the storage program absorbed much of the market shocks.

When carryover stocks were very high, or when the price support was high relative to the world price, the ES curve for the United States had large portions that were flat and other exporting countries were careful to price their product just below the U.S. price so that all of their output would be sold. Bredahl and Green analyzed this issue in the late 1980s.

Deficiency Payment System

In 1986, the United States instituted a second policy to complement the price support system: the deficiency payment scheme. Here the U.S. government guaranteed that producers would receive a fixed target price (P_T) each year, but market prices were allowed to be determined by supply and demand (the government would not purchase grain or encourage storage). Any difference between the market price and the target price was made up by a government payment directly to producers. Producers would make decisions based on the target price, while consumers (domestic and international) would make decisions based on the market price.

Figure 4.9 shows the deficiency payment scheme when the country is a large exporter. The left panel is the exporting country's domestic supply and demand situation. The effective supply curve is vertical below P_T (because producers know they will receive at least the target price), while the entire domestic demand curve is relevant (D_d is the effective demand curve). Exports are equal to the difference between domestic supply and demand, so exports increase under a deficiency payment scheme. If the world price is at P_W' with a deficiency payment program, the country will produce Q_P, consume Q_C, and export the difference. In a pure deficiency payment scheme, there are no stock buildups or trouble in selling internationally because there is no support price that is above the world price.

The right panel of Figure 4.9 shows the excess supply curve for the exporting country. It is kinked at P_T because at prices below P_T there are no supply effects from a lower world price (producers in the exporting country disregard P_W because they are guaranteed P_T). Figure 4.9 shows what the world price and quantity traded would be without the deficiency payment (P_W and Q_T, respectively). If the country is large (as the United States is in most grain markets), the deficiency payment scheme results in a lower world price (P_W') and larger volume of world trade (Q_T').

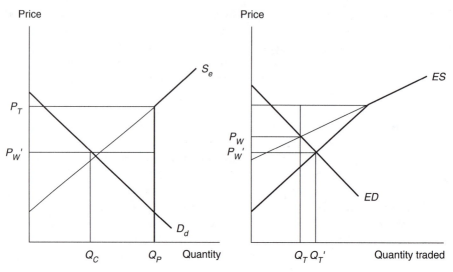

FIGURE 4.9
Deficiency payment system for a large exporting country.

Marketing Boards

The exporting country's counterpart to state trading is the marketing board, which is a government organization that controls exports of a product. As with state trading, marketing boards can allow implicit tariffs, subsidies, or quotas on exports through their operations. There are three general types of marketing boards (Bray et al.): negotiating agencies, which control access to the exporting country's product; central selling agencies, which own the exporting country's product so that all purchasing for export must be done through them; and price and volume regulating agencies, which control prices, production, and all aspects of policy for the exporting country's product.

The government wants these activities performed by a marketing board to influence prices received by producers, reduce price fluctuations, and standardize the terms of sale. If the marketing board purchases the product below world prices and exports the product, there is an implied tax on producers, which is a common occurrence in less developed countries. If the marketing board purchases the product above the world price and exports the product, there is an implied subsidy. Often, though, the marketing board only wants to smooth out prices over the years so that producers do not face wildly fluctuating prices on a yearly basis. A potential problem, however, is that secular downward prices can cause the implied subsidy to increase each year.

The Canadian Wheat Board (CWB) is a good example of a marketing board that has all three of the above-mentioned objectives. By law, all Canadian exports of wheat, barley, and durum must come from the CWB. The CWB is also the sole seller of these products for human consumption. The board dates back to 1917, but it has been the sole marketing agent since 1943 for wheat and since 1949 for other products.

The CWB wants to move all Canadian grain at world market prices without holding government stocks. The Canadian grain elevator and transportation system has limited capacity, so it forms a major challenge for the CWB. The CWB has used rail subsidies (to the West Coast Canadian ports) to make sure that grain flows to export points in an orderly manner. The CWB does not handle the grain itself but allows grain to enter into the marketing system slowly through the quota system it has with farmers. Each farmer has a quota, and the CWB calls for delivery of the quota throughout the year as sales commitments are made.

The CWB follows a pooled pricing system. Farmers receive an initial price upon delivery, and that price (normally about 70 percent of the world price) is essentially the price guarantee. The farmer may receive other payments throughout the year based on the prices the CWB receives. Each Canadian farmer receives the same net price for each grade and class of grain, after accounting for the transportation costs they incur to get the grain to the terminal elevator.

There is a debate in Canada as to whether the CWB is less efficient than a free market system, such as in the United States. Some argue that the CWB has been very flexible in its marketing decisions and has generated tremendous benefits for Canadian grain farmers. According to Schmitz and associates, studies have consistently found that the CWB obtains price premiums over a multiseller arrangement (such as in the United States) because the CWB price discriminates. The Export Enhancement Program of the United States and the European Union export restitutions (both are subsidy schemes) allow Canada to price discriminate among destinations for their wheat, charging high prices where U.S. and E.U. wheat is not subsidized.

Schmitz and colleagues also argue that there is a price premium on Canadian wheat because of its focus on quality throughout the production and marketing system. Government involvement in the marketing system allows the control of quality and more assurance in transactions. Some observers argue that the U.S. grain marketing system has an incentive for companies to market lower-quality grain or at least to make sure that the grain falls within the lower segment of a quality grade. This is not true with the CWB.

SUMMARY

1. Export subsidies result in net losses to the country (losses to consumers and the government, while producers gain). The losses increase if the country is large because the increased exports lower the world price.
2. Export taxes increase welfare to consumers and increase tax revenue to the government, but result in losses for producers. In the small country case there are net losses to the country. It is possible that the country will be a net gainer if it is large because the country restricts exports and increases the world price.
3. If the country is large and can segment its markets, it will maximize its welfare by selling in each market until the marginal revenue in exporting (MRE) is equal

to its marginal cost. The country would charge a high price in markets where demand is relatively inelastic and a lower price in more elastic markets.

4. Price support programs and deficiency payment schemes change the shape of the effective export supply curve for the exporter. The price support program makes the effective export supply curve flat in two places, whereas the deficiency payment scheme kinks the effective export supply curve at the target price.

QUESTIONS

1. Countries clearly lose when they subsidize exports. Why does the United States continue to subsidize exports of some of its products?
2. Export taxes are unconstitutional in the United States. Why do you think that is?
3. Can you think of instances where price discrimination is allowed in the United States? What examples are there in international trade?
4. Should the United States institute a wheat board similar to the one in Canada?
5. How would an export quota be integrated into a supply–demand graph? What would the effective supply curve be?

REFERENCES

Bray, C., Philip Paarlberg, and Forrest Holland. "The Implications of Establishing a U.S. Wheat Board." Foreign Agricultural Economic Report Number 163. International Economics Division, Economics and Statistical Service, U.S. Department of Agriculture, April 1981.

Bredahl, Maury, and Leonardo Green. "Residual Supplier Model of Coarse Grain Trade." *American Journal of Agricultural Economics* 65 (1983): 785–90.

Schmitz, Andrew, Hartley Furtan, Harvey Brooks, and Richard Gray. "The Canadian Wheat Board: How Well Has It Performed?" *Choices.* (First Quarter, 1997): 36–42.

Tweeten, Luther. *Agricultural Trade: Principles and Policies.* Boulder, CO: Westview Press, 1992.

U.S. Department of Agriculture. *Embargoes, Surplus Disposal, and U.S. Agriculture.* Agricultural Economics Report No. 564. Washington, DC: Economic Research Service, 1986.

Chapter 5

Technical Barriers to Trade

Technical regulations and rules (or barriers) are imposed on imports to make sure that international trade does not spread pests, diseases, and other problems to the importing country, and to ensure that imported products meet the same standards required of domestic products. These technical standards include rules for quality, packaging, labeling, standards of identity, and conformity assessment. Some of these regulations help increase the information flow in the marketing process and allow consumers to be informed of product origin, safety, and quality. Other technical regulations or barriers include sanitary and phytosanitary (SPS) regulations on plants and animals to ensure that traded products are not infected with harmful pests or diseases.

New technical regulations on international trade are increasing because of growing demands for enhanced food safety and for an environment that is free from pests and diseases. Consumers want assurances that imported food passes the same health and safety standards that apply to domestic foods. It is natural that as domestic food standards increase, as a result of heightened consumer awareness and enhanced detection technology, so do imported food standards.

This chapter covers general issues associated with technical barriers (their justification, effects, and problems) and specifically looks at how technical barriers were handled in the Uruguay Round of the GATT negotiations. Specific examples of disputes involving technical barriers lend insight into the many problems (mostly political) that come about with liberalization of barriers. The chapter concludes with details on ways of measuring trade barriers quantitatively.

GENERAL ISSUES

There have been long-standing concerns that imported products could harbor harmful nonindigenous species of pests and diseases that could present problems for domestic agricultural industries. Some examples of harmful nonindigenous species that have been brought to the United States include aphids, gypsy moths, and fire ants. It is estimated that 50 to 75 percent of the weeds in the United States and 40 percent of the pests are nonindigenous (OTA). Thus, countries have a right to protect their domestic industries through technical barriers, and they are allowed under the GATT section 20 regulations on health and safety.

Yet, there are concerns that many of these technical barriers are really disguised means of protectionism, brought about because there are limited opportunities for other GATT-legal tariffs and nontariff barriers. Some observers argue that governments are relying on technical barriers to keep imports out and domestic industries more profitable (Orden and Roberts). There is no question that product standards can affect the cost competitiveness of imported products, so it is important that these standards be justified or imported products will be at a disadvantage. The possibility that technical barriers are disguised economic barriers has led to calls for a more scientific approach to justifying and establishing technical standards.

In dealing with technical standards, however, one must also deal with risk, because it is virtually impossible to eliminate the chance that undesired outcomes will occur.[1] Yet there are ways to quantify the risk associated with a particular occurrence and develop a system to ensure that the risk of a bad occurrence is less than some acceptable level. Science can provide quantitative assessment of risks, but it cannot provide answers on acceptable levels of safety. Acceptable levels of risk are normative issues that often come from society or the government. Very low tolerance of risk will generate a demand for very strict technical barriers, while higher tolerances will generate looser technical barriers.

The GATT anticipated problems with the establishment and operations of technical barriers to trade in the 1960s. During the Tokyo Round of GATT negotiations (1974–1979), the Technical Barriers to Trade (TBT) agreement was enacted, which aimed at protecting the consumer against deception and fraud in product standards. The TBT agreement specified that all technical standards and regulations must have a legitimate purpose (to provide information on the product or protect against harmful pests or diseases). The cost of implementing the standard must be proportional to the purpose of the standard, meaning that the costs of implementing the standard cannot be many times the value of the standard in terms of health and safety protection. Finally, the agreement stated that if there were multiple ways of attaining the objective, the least trade-restricting method should be chosen. These are all familiar GATT ideas and expressions that are based on the founding principles of liberalized trade.

The TBT agreement from the Tokyo Round was an important first step, but there was a general consensus that the agreement failed to stem the proliferation of technical regulations (Roberts). The pressure to protect domestic agricultural industries, more scientific findings related to health and food safety, and more accurate detection technology have helped to bring about more barriers and more disputes associated with those barriers. There was not enough legal force behind the GATT dispute settlement procedures before the Uruguay Round, and there was no structured mechanism to reach agreement on legitimate standards and regulations.

---

[1]For instance, it is impossible to keep all weed seeds out of grains. However, one can establish acceptable limits on weed seeds so that an outbreak is very unlikely to occur.

URUGUAY ROUND

The Uruguay Round's Agreement on the Application of Sanitary and Phytosanitary Measures (the SPS agreement) defined basic GATT principles behind use of SPS barriers to trade. The first principle is that countries have the basic right to institute SPS measures that protect plant, animal, and human health when they are based on scientific principles. These measures cannot discriminate between members where identical or similar conditions prevail. If the European Union and the United States have two different scientifically valid means of ensuring that milk is pure, the United States cannot insist that the E.U. follow its procedure. The United States must allow imports from the E.U. that have followed the E.U.'s scientifically valid methods.

A second principle is that members should base their SPS requirements on international standards, guidelines, or recommendations when they exist, although they can institute higher standards if there is a scientific justification. This implies a need to harmonize SPS measures among countries so that international trade will flow more easily and production processes can be geared toward one set of regulations. New Zealand is a major exporting nation for many agricultural products. Its beef industry must conform to standards in the E.U. and the United States; its apple industry must conform to standards for countries throughout the world with regard to the 146 organisms that affect mature fruit (Johnson). A common world standard for SPS and other TBTs would be helpful for exporters.

A third principle is that members are obliged to recognize measures adopted by other countries that provide equivalent levels of protection. This is the concept of equivalency and means that countries should view the end product rather than the process that is used to meet the standards. The Japanese cannot reject a fumigation method used by the United States simply because it is not used in Japan. If the method is deemed safe and effective on a scientific basis, the Japanese must recognize it as equivalent.

A fourth principle is that members should base their SPS measures on a risk assessment, taking into account methodologies developed under the auspices of three international organizations: the Codex Alimenetarius Commission (Codex), the International Office of Epizootics (IOE), and the Secretariat of the International Plant Protection Convention (IPPC). These organizations will help in moving the world toward a common risk assessment methodology and harmonized standards.

The final principle is that members should recognize that SPS risks do not necessarily coincide with political boundaries. Pest-free and disease-free areas within a country should be recognized if there is sufficient evidence to warrant. Therefore, import protocols should be based on a regional distinction in many cases, rather than a country distinction. This principle will be important in establishing pest-free and disease-free zones within countries from which exportation will be allowed.

These principles, along with the enhanced dispute settlement procedures, should help the GATT and its successor agency, the WTO, to ensure that SPS and other technical barriers are truly in place to protect health and safety. Another important part of the SPS agreement requires countries to provide advance notification for standard changes. There were 400 such notifications within the first eighteen months of the WTO's existence (Stanton).

The risk assessment methodology for SPS measures suggested by the GATT involves three steps: evaluating the likelihood of a disease or pest entering a country or determining the potential adverse effects on health of additives and contaminants (the scientific part of the process); determining the acceptable level of risk that can be tolerated (this is a choice that decision makers must justify because rarely is there zero chance of occurrence); and selecting and applying measures that would limit risk to acceptable levels (designing a procedure that is as trade-neutral as possible).

If two parties disagree on a particular technical barrier, the WTO dispute settlement procedure begins with initial consultations among the parties. If the initial talks do not solve the dispute, a party can request that a WTO dispute panel be formed. The panel will hear the case and rule on it. The losing party is obliged to implement the panel's recommendations and report how it has complied. This is a very standardized process, and the decisions of the panel have more status than under the GATT rules. The international organizations referenced earlier will provide technical expertise in the dispute settlement process.

These three international organizations—Codex, IOE, and IPPC—are well recognized and have a long history of issues on health and safety. Codex, established in 1963, is responsible for food additives, pesticide residues, contaminants, animal drugs, packaging, and food standards. It is a subsidiary of the Food and Agriculture Organization and the World Health Organization. Its 153 member countries cover 98 percent of the world's population with representatives from government regulatory agencies, the scientific community, and the food industry (Dawson). It works with member countries to establish food standards to facilitate international trade.

The International Office of Epizootics (IOE), an international veterinary organization that was formed in 1924, is responsible for animal health issues. There are 130 countries represented in its membership, and the organization maintains a reporting network on animal diseases. The IOE would be useful in establishing quarantine policies for livestock trade and determining the procedures to decide whether a region is disease-free (which is a big issue for foot-and-mouth disease).

The International Plant Protection Convention (IPPC), which was formed in 1951, is responsible for plant disease and plant health issues. It is another subsidiary of the Food and Agriculture Organization, and it has representatives from around ninety countries.

All of these international organizations are very active in developing approaches to manage risk and working with individual governments to recommend quality control programs. Codex has been active in working on hazard analysis critical control point (HACCP) and best management practices (BMPs) systems. Their involvement ensures that no country can dominate discussions on technical barriers and that science will play the key role in decisions.

Another important international organization for the food industry is the International Standards Organization, located in Geneva, Switzerland. Its ISO 9000 program is a voluntary quality assurance system to certify that the best practices are being followed by firms. This internationally recognized standard helps importers know that the product has reached a process criteria that is appropriate for the particular industry.

It would be nice if the world could agree on common standards, publish and abide by them, and allow firms to trade freely based on those agreed-upon standards. However, that is not exactly how the world works even among developed countries—but it is moving in that direction. The costs involved in establishing standards have increased greatly over the years, and science is more able to provide precise product quality measurements. However, tastes and situations vary tremendously by country and there are preexisting national laws in force. As product standards increase in the future (as a result of higher incomes and increased preference for food safety), countries with reliable food safety and animal and plant health systems will have an advantage over others.

EXAMPLES OF DISPUTES OVER SPS MEASURES

This section covers three SPS cases that highlight the difficulty in settling disputes between and among countries. Many times politics plays a more pronounced role than science. Yet there are clear indications that over time the WTO mechanisms will allow science to prevail, even for large, powerful countries.

Mexican Avocados

Mexican avocados have been banned from entering the United States since 1914 because of a fear that their importation would lead to infestations of avocado seed weevils (moths and fruit flies) for American producers.[2] The Mexican avocado industry is concentrated in the southwestern part of the country, and the mountains of northern Mexico have blocked the weevil's northern movement. Mexican scientists claim that there are modern chemicals and cultural practices that can be used to quell the threat.

Mexico is the largest avocado exporter in the world, and it would love to reach the U.S. market for avocados. A study commissioned by the American Farm Bureau estimated that Mexican production costs were $600 to $900 per acre compared with $5,200 to $5,700 per acre in California. Roberts and Orden, when comparing wholesale prices between Mexico and the United States, found that U.S. prices were approximately double those in Mexico. There is no question that Mexico is a low-cost producer of avocados and that their produce would flood into the United States if there were no technical and trade barriers.

The negotiations for NAFTA prompted a new inquiry by the Animal Plant Health Inspection Service (APHIS) of the U.S. Department of Agriculture, which is the organization in charge of quarantine issues associated with international trade, into the import situation for Mexican avocados. APHIS assesses hazards and recommends rules that permit entry of imported products with minimal risk to U.S. agriculture. This review began in 1990 and resulted in a proposed rule by APHIS to allow importation of avocados from the Mexican state of Michoacán, which is in northern Mexico,

[2]This case is based on a study by Roberts and Orden.

into the northeastern United States during the months of November through February. These months were chosen because the weather conditions minimized the risk of pest infestations. Field studies had detected no weevils in Michoacán. The APHIS ruling also established strict regulations for monitoring insect populations, harvesting, packing and shipping practices, and inspections.

The review period for the proposed regulation change and the public comment period were spiced with negative comments from the U.S. avocado industry. The U.S. industry argued that the field studies on insects in Michoacán were flawed and that the United States should allow importation of Mexican avocados only if all regions of Mexico were deemed pest-free. The final proposed rules reflected numerous changes in the work plans that the Mexican authorities had devised for the production and importation process. In the final analysis, APHIS estimated that a seed or fruit fly outbreak might occur less than once every million years and a stem weevil outbreak once every 11,402 years.

It is interesting to note that other U.S. fruit growers were not sympathetic to the cause of the U.S. avocado industry. A pest infestation could cause severe damage to other U.S. fruit growers, but they did not object to allowing importation of Mexican avocados. They saw the entire strategy of the U.S. avocado industry as seeking protection to keep avocados out of the United States, which harmed efforts to open Mexican fruit markets to increased competition. Roberts and Orden imply that there was tremendous pressure on APHIS to keep the import ban on Mexican avocados, despite a minimal risk of pest infestation.

Nursery Imports

Romano and Orden provide another case where the U.S. industry has put pressure on the U.S. Department of Agriculture, specifically APHIS, to keep import restrictions that are not justified by scientific evidence. The nursery case has been ongoing since the early 1970s, and it is illustrative of the longtime delays and meticulous negotiations that often accompany procedures to reduce technical barriers.

The importation of most nursery stock, plants, roots, bulbs, and other plant products is allowed only in bare-root condition because the attached soil can carry undesirable pathogens. Shipping plants on a bare-root basis is very costly because of high mortality rates. In 1974, the Netherlands Inspection Service reached an agreement with APHIS that allowed five plant genera to enter the United States in soilless growing media, including unused peat moss and vermiculite. These shipments were allowed on a five-year trial basis under the stipulation that strict phytosanitary procedures be followed. There was little at stake for the U.S. nursery industry because the value of imports was quite small for these five items. No disease outbreak occurred during the five-year trial period, so APHIS ruled that importation of these genera could occur with the specified growing media.

Between 1980 and 1983, APHIS received requests to allow importation of an additional sixty genera in the same growing media. Because of the comments and pressure from the nursery industry (probably related to the economic significance of these genera), APHIS stopped the regulatory process during the middle 1980s. Ultimately, APHIS decided to perform the risk assessments (the major part of the regulatory process) on five products at a time, in the hope that this would defuse some of the controversy associated with the economic impacts. Throughout the process there has been a struggle between APHIS scientists, who focused on minimal risk strategies (consistent with the GATT), and the industry, which wanted a zero risk strategy (obtainable only through an import ban). To date, only ten of the sixty requests have been approved.

Romano and Orden argue that this is another instance where interest groups have exerted strong pressure on the regulating agency to use SPS measures as a means to discourage international trade. There is no scientific evidence that gives credence to the U.S. nursery industry's viewpoint. These types of disputes involving risk assessments are detrimental to the trade liberalization process and make it impossible to arrive at a consensus among the disputing parties. Both sides can agree on the probability of an outbreak, but if one side demands a minimal risk and the other side demands no risk, there is no room for imports.

Beef Treated with Growth Hormones

The European Union's (at the time, it was called the European Community's) Council of Ministers adopted a ban on the use of growth-promoting hormones in domestic cattle in 1985 and banned imports of cattle and beef that had been treated with growth hormones in 1988. Much of the stimulus for this ban on growth hormones came from E.U. consumers, who had experienced a number of food safety scares dating back to a scandal in Italy where the illegal growth promotant DES was found in baby food. Europeans have been skeptical of growth hormones since that time.

The United States viewed the hormone ban as simply a protectionist measure to keep U.S. beef out of the European Union. The ban occurred at a time when there was tremendous pressure on the E.U. to reduce agricultural expenditures. The E.U. had huge surpluses of beef and other animal products, so the United States felt that the Europeans were trying to keep U.S. beef out of the market to reduce their surplus beef stocks.

The United States tried to use the TBT agreement from the Tokyo Round to attack the hormone ban in 1986 (before the import ban), but the TBT agreement dealt only with end-product characteristics. One cannot tell whether beef has been treated with a growth-promoting hormone because such hormones occur naturally in beef. The hormone ban involved restrictions on a specific production method, so the United States tried unilateral negotiations for a number of years, but they did not work. At the same time a Codex Commission decided that there is no need to place a maximum residual limit on natural hormones (the type used in the United States during the cattle production process).

Finally, the United States and Canada began the WTO dispute settlement procedure with the E.U. in 1996. A panel was formed, and arguments were heard from both sides. The E.U. stated that its regulations on hormone use were a precautionary stance because one never knows what the ultimate risks will be. It argued that it had a right to impose a higher standard to support the concerns of its citizens.

The United States countered with a challenge that there was no scientific evidence that growth-promoting hormones cause health problems. The existence of hormones in food is not related to health, since one egg has over seventy-five times the amount of hormones in 1 kilogram of beef. One cannot simply argue that banning hormone use is a higher standard without supporting evidence. Further, the E.U. allows similar hormones in other situations (such as carbadox in swine production), so there was significant evidence that the ban was a protectionist scheme.

The WTO panel rendered its decision in 1997 and sided with the United States and Canada. It said there was no proof that disallowing hormone-treated beef was indeed a higher standard. The panel suggested that the E.U. develop a labeling program to identify hormone-treated beef, so that if consumers wanted such beef, they could readily purchase it.

The E.U. notified the WTO that it wanted to appeal the panel's decision, but the Appellate Body also ruled in favor of the United States and Canada. After the decision was adopted by the WTO Dispute Settlement Body, the E.U. said it would abide by the decision in "a reasonable time." An arbitrator decided that a reasonable time was fifteen months, so the E.U. had until May 1999 to comply with the WTO ruling. The E.U. did not abide by the WTO decision so the United States was allowed to withdraw access for $117 million in E.U. products.

Other trade issues associated with hormones are likely to occur in the future. Some people would like to ban milk from dairy cattle treated with recombinant bovine somatotropin (rbST). Again, this is a production process standard, and one cannot distinguish milk from cattle treated with rbST. The U.S. Food and Drug Administration will not allow milk processors to identify such milk because it is identical with milk from untreated dairy animals. Yet, it is likely that the E.U. will have a different view on this issue, especially when its consumers do not want milk or dairy products from cattle treated with rbST.

FUTURE CHALLENGES FOR THE WTO

There has been great progress in dismantling some SPS barriers in recent years. Seven formal complaints were filed within the first eighteen months after the WTO came into existence (Stanton). These complaints cover inspection procedures for fresh fruit, shelf-life requirements for processed meats, treatment requirements for bottled water, diseases associated with fish, and hormone treatment of meat. There are also disputes over shelf-life requirements for foods, irradiation of foods, nutrition labeling, and other issues.

Progress is being made, but it is slow because technical requirements must be addressed on a case-by-case basis with much effort going into the risk assessment methods. The U.S. Department of Agriculture's Foreign Agricultural Service estimates that there are 292 questionable foreign technical measures that impact U.S. agricultural exports, reducing them by more than $5.4 billion (Roberts). Nonetheless, as methodologies become standardized and countries become convinced that WTO dispute settlement procedures are working, there will be more unilateral or bilateral resolution of disputes, resulting in fewer technical barriers to trade.

DISCUSSION BOX 5.1

European Consumer Fears about Synthetic Agricultural Products

The beef-hormone issue is indicative of a much larger fear in Europe concerning modification of agricultural products. There have been cases where European scientists have insisted that production processes were safe, only to discover later that they were not. The most controversial, recent example of this is the problems with bovine spongiform encephalopathy (BSE), or mad cow disease. British scientists had stated that the feeding practices of farmers were safe and would not affect people consuming British beef. However, when it was discovered that mad cow disease could be transmitted to humans consuming the beef, European consumers wondered what scientific revelations would come for other food products.

A major fight is brewing now over the use of genetically modified organisms (GMOs). Many crops in the United States emanate from genetically modified seed stock, and some countries, including many in Europe, want these GMOs separated from natural products and labeled as GMO. This separation of product would be nearly impossible within the current grain marketing system, which means that grain sellers cannot guarantee that a shipment is GMO-free. The United States shipped 9 million metric tons of soybeans to the European Union in 1998, and all of those beans were genetically modified.

The Europeans are not alone in their concern. Many African and Latin American nations are in favor of labeling programs for GMO products. However, most grain exporting nations—including Argentina, Australia, Canada, and the United States—are vehemently opposed to labeling. There is no scientific evidence that GMO products cause harm to consumers, though a recent British study, which has been judged as nonscientific, found that genetically modified potatoes caused stunted growth and suppressed immunity in laboratory rats.

Labeling proponents argue that one never knows whether future research will find GMO products harmful. There is a history of this in Europe, so why not label the products and allow consumers to choose? This isn't any different from labeling meats as kosher/halal or tuna as dolphin-friendly: the labeling provides information to the consumer. A recent survey in Europe found that 86 percent support labeling of GMO products (*The Economist*). Yet producers are worried that nonscientific consumer concerns will keep them from reaping the benefits of GMO products and force them to identify their products for many markets (increasing their costs tremendously).

MEASURING TRADE BARRIERS

As this chapter and previous ones have shown, there are a host of different methods for countries to break the link between their domestic market and the world market. The break can result from a tariff, quota, variable levy, or technical requirements. Often there are numerous policies, particularly with technical requirements, that impact the linkage between the domestic market and the world market for a particular product. It is essential to measure the protection that these trade barriers afford, so that trade negotiators can understand what barriers should be negotiated away and liberalization progress can be measured as barriers are lifted.

This section covers three common measures of trade barriers used for monitoring trade liberalization: the nominal rate of protection, the effective rate of protection, and the producer (and consumer) subsidy equivalent.

Nominal Rate of Protection

It is possible that one could quantify the protection that each trade barrier affords by investigating the effect that the barrier has on the difference between the world price and the domestic price. A $1 per unit tariff would generate a $1 difference between the world price and the domestic price. However, this only works with barriers that have prescribed price effects, such as tariffs. As previous chapters have shown, the domestic price effects of a quota will depend on supply and demand conditions within the country. Measuring the price effects of technical requirements and other restrictions that are not price-based is virtually impossible.

For this reason, economists tend to look at the difference between domestic prices and world prices to determine the combined effects of all trade barriers for a particular product. One must be careful that the two prices are comparable (at the same level within the marketing system and the same geographic location) so that the price difference reflects only the barriers to trade (a border price compared with an internal wholesale price at the port). Thus, the difference between the domestic price, P_d, and the world price, P_w, is the implied trade barrier, T.[3]

$$T = P_d - P_w \qquad (5.1)$$

For example, if the product in question is peanut butter, the world price might be $1.50 per pound, while the domestic price might be $2.00 per pound, making the implied trade barrier $0.50 per pound.

The nominal rate of protection, NRP, is the simplest percentage measure of trade barriers. It simply puts Equation (5.1) into percentage terms.

$$NRP = T/P_w \qquad (5.2)$$

In the peanut butter case, the NRP would be 33 percent (0.50/1.50).

[3]Note that T will include any taxes or subsidies that are applied exclusively to the domestic product. If there are taxes that apply only to the domestic product, it is possible that T would be negative.

Effective Rate of Protection

If one wishes to measure the degree that domestic producers of a product are protected, then consider that the value-added by the manufacturer is less than the product's price. Assuming that the price of the processed product is broken into the price of the raw material (input 1) and the composite of all other value-added activity (input 2, which would include land, labor, capital, etc.), then the following equation holds in perfect competition:

$$P_d = w_1 P_1 + w_2 P_2 \tag{5.3}$$

One can think of the w_is as reflecting the number of units of input i needed to produce 1 unit of the processed good. The value-added of the processor is $w_2 P_2$, because $w_1 P_1$ is a purchased input. The effective rate of protection, ERP, reflects only the protection afforded the value-added by the processor:

$$ERP = T/(w_2 P_2) \tag{5.4}$$

or

$$ERP = (P_d - P_w)/(w_2 P_2) \tag{5.5}$$

Note that the ERP will normally be greater than or equal to NRP. The lower the value-added in the process, the higher the effective rate of protection. In the peanut butter case, if one assumes that peanuts are the only raw ingredient in peanut butter and those peanuts make up 50 percent of the domestic cost, the ERP for peanut butter would be 50 percent (0.50/1.00). If the costs of peanuts make up only 10 percent of the cost, the ERP would be 27.8 percent (0.50/1.80), which is lower than the NRP because the value-added is quite high for peanut butter manufacturers.

To further complicate things (but to get closer to reality), assume that the processed product uses an ingredient whose domestic price is above the world price (i.e., there is an implied trade barrier on the ingredient). The protection afforded domestic producers of the processed product is lower because they must pay a higher price for their input.

The price of the processed product without trade barriers on either the product itself or the ingredient is reflected in Equation (5.3) if P_1 reflects the world price for the ingredient. However, because of trade barriers on input 1, (the domestic price for the input is $P_1{}^*$), the domestic price of the processed product under perfect competition with all trade barriers, $P_d{}^*$, is:

$$P_d{}^* = w_1 P_1{}^* + w_2 P_2 \tag{5.6}$$

The difference between the domestic price of the processed product, $P_d{}^*$, and what it should be under free trade, P_d, has two elements—an implied trade barrier on the processed product, T^*, and an implied trade barrier on input 1, $w_1 (P_1{}^* - P_1)$.

The implied trade barrier on the processed product can be found from the following equation:

$$P_d - P_w = w_1 (P_1{}^* - P_1) + T^* \qquad (5.7)$$

or

$$T^* = P_d - P_w - w_1 (P_1{}^* - P_1) \qquad (5.8)$$

In this case, the implied trade barrier, T^*, is less than the difference between the domestic price and world price for the processed product because the processor has purchased inputs at above the world price.[4] The formula for calculating the ERP is still Equation (5.4), but with T^* instead of T.

In the peanut butter example, if the domestic price of peanuts was double the world price, so that the world price of peanuts in 1 pound of peanut butter was only $0.50, then the effective rate of protection on peanut butter would be zero. The peanut processor is paying $0.50 extra for the peanuts in each pound of peanut butter, and the domestic peanut butter price is $0.50 above the world price. In this example, the trade barrier on peanut butter has fully compensated the peanut butter producer for the higher price of domestic peanuts.

Producer and Consumer Subsidy Equivalents

Nominal and effective rates of protection are good measures on an individual product basis. However, they are not perfect, particularly for agricultural products, because of the important role of government subsidies. Some of these subsidies come in the form of direct payments to producers, which are easy to include in an NRP or ERP formula, but many areas of government support are not directed at a particular commodity or are difficult to document in terms of a difference between a domestic price and a world price. Producer subsidy equivalents (PSEs) were developed to overcome some of the problems with NRPs and ERPs.

The PSE measures the effects of government policy on gross farm income, or the lump sum payment that governments would need to pay farmers to keep their income unchanged if government policies terminated (Ballenger). PSEs include direct government payments (or subsidies) plus the costs of government expenditures related to the commodity. PSEs are normally calculated by the following general formula:

$$PSE = (P_d - P_w) Q + GE \qquad (5.9)$$

The PSE can also be shown as a percentage of farm receipts or on a per ton basis.

---

[4]Note that $P_1{}^*$ can include government taxes or subsidies that cause the domestic price of the input to vary from the world price.

TABLE 5.1   Calculation of PSEs for U.S. Corn, 1986 and 1993*

	1986	1993
Direct payments to producers		
Crop insurance	(17)	0
Deficiency payments	6,196	1,526
Disaster payments	0	0
Diversion payments	133	0
Loan forfeit benefits	1,333	0
Storage payments	359	0
Total direct payments	8,004	1,526
Indirect transfers		
Commodity loans	879	0
Farm credit	208	135
Pest and disease control	27	43
Advisory	33	39
Inspection	5	7
Processing and marketing	13	17
Land improvements	150	201
Research	84	128
State programs	200	249
Taxation	132	53
Transport	57	122
Total indirect transfers	1,788	944
Total direct payments and indirect transfers (PSE)	9,792	2,520
Value of production	12,507	16,597
Direct payments	8,004	1,526
Total income	20,511	18,123
PSE as a percentage of income	47.7%	13.9%

*In millions.
Source: Economic Research Service, U.S. Department of Agriculture.
http:www.econ.ag.gov/Prodsrvs/dp-ti.htm#subsidy

The term $P_d - P_w$ is the price wedge between domestic prices and international prices, Q is the amount of production, and GE includes government expenditures that also support producers of the product, including general and product-specific expenditures. When PSE is divided by gross farm income, it is a percentage of gross farm income that is attributed to government policy.

Government expenditures that are not associated with particular commodities (such as farm credit subsidies, transportation subsidies, and government expenditures on research and extension) are normally assigned to commodities in their proportion of farm income (e.g., if corn accounts for 10 percent of farm income, 10 percent of the general government expenditures are included in PSE calculations for corn).

Table 5.1 shows PSE calculations for corn in the United States for 1986, the year where the PSE as a percentage of corn producer income was at a maximum, and 1993, the latest year available. For 1986, direct payments to corn producers totaled $8.0 billion and indirect transfers totaled $1.8 billion. These benefits to corn producers accounted for 47.7 percent of their total income. The PSE for 1993 was much lower because direct payments to corn producers were $1.5 billion and indirect transfers were $0.9 billion. These benefits to corn producers accounted for 13.9 percent of their total income. There was no difference between the U.S. corn price and the world corn price, so the price wedge was zero for both years.

One benefit of the PSE is that it is readily comparable between countries because it includes the scale of production and involves a dollar amount. Further, one can easily aggregate PSEs for commodities to obtain a PSE for the entire agricultural industry of a country. This is impossible to do with NRPs or ERPs without tremendous weighting problems by product. So PSEs provide a simplified yardstick to measure one country or commodity against another.

PSEs are also convenient ways of turning every barrier and government expenditure into a subsidy equivalent. Pure NRPs and ERPs do not consider whether domestic production is great or small, so an NRP of 10 percent on beef (a major import item for the United States) has the same presumed impact as an NRP of 10 percent on corn (a product where U.S. imports are virtually zero). The subsidy equivalent of the 10 percent NRP on beef will far exceed the subsidy equivalent of the 10 percent NRP on corn for the United States. When multilateral negotiations are taking place, it is far easier to deal with PSEs than with other measures of protection. Also, when the negotiations are finished, dealing with PSEs makes it easier for countries to arrive at liberalization targets: it doesn't matter what combination of policies is changed, only that the PSEs fall by a certain percentage.

Finally, the PSE serves as a convenient monitoring device because it is relatively easy to calculate (though time-consuming). In the Uruguay Round of the GATT negotiations, the final negotiated outcome by country involved aggregate PSEs and PSEs by individual commodity, so having measures that are easy to calculate (and negotiate) is very important.

Despite their ease of use and encompassing definition to include all government transfers, PSEs have their problems with respect to trade liberalization. They do not indicate the impact of government policies on world markets. Specifically, PSEs do not measure the degree that international trade has been distorted, and elimination of this distortion is the main topic of multilateral trade negotiations. Two countries can have identical PSEs for a commodity, but one country's policies might have no impact on trade while the other's policies might have tremendous impacts.

Consider a case where an exporting country guarantees the price of the product above the world price, as depicted in Figure 5.1. With no government policies and free trade, the country should allow its domestic price to equal the world price, P_w, and exports would be Q_1 minus Q_2 (production minus consumption). Assume that the country instead wants to guarantee its producers a price of P_s through a deficiency payment that would make up the difference between P_s and the market price (which would be P_w since the country is an exporter). However, producers are only allowed to receive the price guarantee if they set aside acreage (and assume that percentage is enough to shift the effective supply curve from S to S').

FIGURE 5.1
Combination of policies with no net effect on exports.

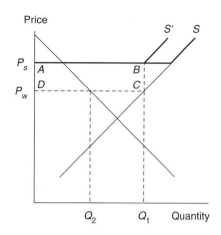

The PSE for the country would be the area *ABCD*, but this combination of policies would have no effect on the country's exports, and therefore no effect on the rest of the world. Other countries shouldn't care about this combination of policies because it does not affect world price, consumption, or production. One can contrast this with a pure export subsidy system generating an identical PSE, which would increase the exporting country's exports and therefore generate trade distortions that would cause the rest of the world to change their production and consumption patterns. Thus, one must be careful in inferring that PSEs always reflect trade distortions.

Calculations of PSEs have problems that they share with other measures of protection. Because world prices and domestic prices are often denominated in different currencies, one must choose which exchange rate to use. This choice can be a problem if exchange controls exist (restrictions on who can exchange currency or uses of the currency) or if there are multiple exchange rates depending on the transaction. Both of these instances occur in some less developed countries. Further, recent volatility in exchange rates and world commodity prices can change the calculations drastically over time. Averaging exchange rates and commodity prices over time may reduce the variability of these calculations.

Despite problems with measuring trade barriers, there is a huge demand for aggregate measures that can be used for negotiations and monitoring progress in trade liberalization. It is important that work continue to refine these measures and overcome some of the conceptual weaknesses, while keeping them relatively simple for policy formation.

Finally, consumer subsidy equivalents are comparable to PSEs, but they measure the value of government subsidies to consumers and are often denoted as a percentage of consumer expenditures. CSEs also have two components: the price wedge between the world price and the domestic price (CSEs are negative if the domestic price is above the world price) and government expenditures that benefit consumers of the commodity. They are typically small for most countries, but can be large positive numbers for some less developed countries.

SUMMARY

1. Technical standards are rules for quality, packaging, labeling, standards of identity, and conformity assessment. Some of these regulations help increase the information flow in the marketing process and allow consumers to be informed of product origin, safety, and quality. Other technical regulations or barriers include sanitary and phytosanitary (SPS) regulations on plants and animals to ensure that traded products are not infected with harmful pests or diseases.

2. Many people are concerned that technical barriers are purely means of protection for domestic producers. This is especially true now when traditional trade barriers have fallen as a result of GATT negotiations, but countries are increasing their technical barriers.

3. The Uruguay Round's SPS agreement had the following basic principles: countries have the right to institute measures that protect plant, animal, and human health if they are based on scientific principles; countries should base their SPS requirements on international standards, guidelines, or recommendations; countries should recognize measures adopted by other countries if they provide equivalent protection; SPS measures should be based on a risk assessment; and countries should recognize that SPS risks do not necessarily coincide with political boundaries.

4. The Mexican avocado, nursery import, and beef-hormone cases show how difficult it is to make trade more open through reducing technical barriers. There are numerous ways to slow the liberalization process, particularly if the money involved is substantial.

5. Four measures of trade barriers are the nominal rate of protection, the effective rate of protection, the producer subsidy equivalent, and the consumer subsidy equivalent. The first two deal with differences between domestic prices and world prices, whereas the latter two also incorporate government expenditures into the calculations.

QUESTIONS

1. How do you feel about the United States harmonizing its food and safety standards to those of an international scientific organization?
2. How would you feel about the Mexican avocado case if you were an avocado producer?
3. Technical barriers to trade are becoming increasingly contentious issues in agricultural trade. How do you think GATT philosophies and dispute settlement affect the implementation of technical barriers?
4. Why were PSEs useful in GATT negotiations?
5. Should products derived from genetically modified organisms be labeled as such? Why?

REFERENCES

Ballenger, Nicole. "PSEs: What They Are and Their Role in Trade Negotiations." *Choices* (First Quarter, 1988): 36–37.

Dawson, Richard. "Impact of WTO on CODEX Alimentarius and Its Implications for World Trade." In Orden and Roberts (Eds.), *Understanding Technical Barriers to Agricultural Trade.* IATRC, pp. 69–74.

The Economist. May 1, 1999, p. 75.

Johnson, R. W. M. "Technical Measures for Meat and Other Products in Pacific Basin Countries." In Orden and Roberts (Eds.), *Understanding Technical Barriers to Agricultural Trade.* IATRC, pp. 79–98.

Office of Technology Assessment (OTA), U.S. Congress. "Agriculture, Trade and Environment: Achieving Complementary Policies," May 1995.

Orden, David, and Donna Roberts. *Understanding Technical Barriers to Agricultural Trade.* Proceedings of a Conference of the International Agricultural Trade Research Consortium, January 1997.

Roberts, Donna. "Implementation of the WTO Agreement on the Application of Sanitary and Phytosanitary Measures: The First Two Years." International Agricultural Trade Research Consortium Working Paper #98-4, May 1998.

———, and David Orden. "Determinants of Technical Barriers to Trade: The Case of U.S. Phytosanitary Restrictions on Avocados, 1972–1995." In Orden and Roberts (Eds.), *Understanding Technical Barriers to Agricultural Trade.* IATRC, pp. 117–160.

Romano, Eduardo, and David Orden. "The Political Economy of U.S. Import Restrictions on Nursery Stock and Ornamental Plants in Growing Media." In Orden and Roberts (Eds.), *Understanding Technical Barriers to Agricultural Trade.* IATRC, pp. 99–116.

Stanton, Gretchen. "Implications of the WTO Agreement on Sanitary and Phytosanitary Measures." In Orden and Roberts (Eds.), *Understanding Technical Barriers to Agricultural Trade.* IATRC, pp. 75–78.

Chapter 6

Multilateral Trade Negotiations: The GATT and WTO

After World War II, the Allies made great efforts to ensure that all countries of the world had forums where they could meet, discuss issues, and decide on the best course for the world to move forward. The most important and encompassing organization formed was the United Nations, which is mostly for political matters and international relations. The idea was that if nations continually communicated in an open forum, it was less likely that conflicts would occur.

Numerous multilateral organizations were also formed to ensure that the world trading system operated smoothly, because many believed that economic matters were a major cause of Germany's belligerent behavior during the 1930s (the high reparations from World War I and resulting inflation allowed German fringe parties to be successful).

This chapter covers the major multilateral organization charged with promoting freer trade among nations: the General Agreement on Tariffs and Trade (GATT). The guiding principle behind the GATT is that trade benefits all countries, as seen in Chapter 2. Yet people realize that domestic issues often encourage countries to follow more restrictive trade policies. The GATT is structured to overcome those domestic arguments for trade barriers through multilateral negotiations involving tariff and other trade barrier reductions among all GATT member nations. Through this mechanism, trade barriers have fallen drastically in most products and world trade has soared, resulting in vastly improved welfare for all countries.

GATT HISTORY AND GUIDELINES

The GATT was established after World War II (in 1947), along with the United Nations, International Monetary Fund, and International Bank for Reconstruction and Development (World Bank). There were twenty-three original GATT members.

Until 1994, when the Uruguay Round ended and the World Trade Organization (WTO) was established (on January 1, 1995), the GATT was the only multilateral organization with established rules for international trade policies. The GATT had member nations (117 in 1993) that pledged to follow the principles of reducing trade barriers and other trade distortions. The members included all major trading countries except China and the former Soviet Union. The GATT, and now the WTO, is mostly a negotiating club with specific rules for conduct and dispute settlement.

The GATT rules and principles are only as good as the pledges and promises of the member nations. There is no international government or police force to uphold the GATT rules and agreements. National sovereignty is always an issue because multilateral rules can have uncomfortable impacts on member nations. Thus the GATT works only as well as the member nations make it work through their own self-interest in lowering trade barriers to promote and increase world trade. There is no question, though, that the GATT rules and principles make it easier for member governments to make changes to open their economies.

The GATT is guided by a number of principles that members agree to uphold. These include:

reciprocity	A country comes to the bargaining table with offers to reduce trade barriers with the expectation that other countries will do the same.
nondiscrimination	Countries are to give all GATT members the same preferences. This is often called the most favored nation principle.
transparency	Trade barriers should be easily recognized by others, not disguised.
national treatment	Goods should receive the same treatment within a country, no matter what their country of origin.
compensation	Countries harmed by changes in policies of another country are entitled to compensation.

All GATT member countries pledge themselves to these principles and the overall goal of freer trade. Countries must convince the GATT, and now the WTO, of their commitment before membership is accepted.

The GATT also allowed for dispute settlement and sanctions if rules were violated. Countries would normally try to work out difficulties themselves if a disagreement erupted, but sometimes they would take their dispute to the GATT. A GATT panel of experts would hear the arguments of both sides and render a verdict. However, if the offending party was found guilty, the system relied on the offending country to remedy the situation: the GATT did not force a policy change. If the offending country did not remedy the situation, the injured country was allowed to enact sanctions on the offending party in proportion to the estimated injury. Yet a major problem was that in the end, the resolution of any dispute relied on the voluntary action of the offending government.

The manner in which these principles were followed during the GATT's history had its ups and downs, depending on how committed the member governments were to the GATT's goals. Nonetheless, there has been tremendous liberalization in merchandise trade since 1947, and certainly the GATT is a major reason for this occurrence.

Trade negotiations under the GATT were generally handled through Rounds (occurring approximately every eight years) that concentrated on trade barriers and other problems in specific industries. This allowed countries to come with very specific proposals on liberalization for their economy, along with certain expectations for liberalization on the part of other countries. The preparation undertaken and effort expended were often great, but the push to end a Round, with the resulting successful liberalizations, made the GATT Round format a success. Further, because countries had pledged never to retrench and enact higher trade barriers, each Round saw fewer factors inhibiting trade over time.

The one exception to this continual reduction in trade barriers through the GATT was agriculture (at the insistence of the United States in 1955). The U.S. Congress forced the GATT to allow waivers on policies that would conflict with U.S. agricultural policies. Other countries agreed because agriculture had always been viewed as a special case. Three exemptions of note were that, first, export subsidies could be used on agricultural products as long as they didn't result in more than an "equitable share" of the market. A second exemption was in section 22, where countries were allowed to enact tariffs if imports were found to cause "domestic injury" for agricultural products covered by a domestic policy. The third exemption allowed suspension of GATT rules if needed to preserve health, safety conservation, and national security.

The nebulous nature of these exemptions and the continuing desire of countries to protect their agricultural sector (especially in Japan, Western Europe, and the United States) effectively kept agricultural products out of the negotiation Rounds until 1986, when the Uruguay Round began. That year, world agricultural prices were at their lowest level in many years and agricultural protections, measured by producer subsidy equivalents (PSEs), were at their peak. The beginnings of these protectionist policies trace back to the establishment of the united Western Europe in 1958 (the European Common Market at that time) and its Common Agricultural Policy (see Chapters 7 and 10). The emergence of the European Community as a major agricultural exporter during the later 1970s and throughout the 1980s encouraged the United States to rethink its idea of keeping agriculture out of the GATT.

The challenges for agriculture came from high levels of internal support, export subsidies that allowed countries to dispose of accumulated surpluses, and high trade barriers to support domestic policies. The high internal supports (usually high guaranteed domestic prices) in many countries encouraged production and exports. This hurt producers in countries that did not subsidize their agriculture, or subsidized it at a relatively low level. Export subsidies were coupled with high internal support, particularly by the European Community, so that surpluses could be dumped on the world market.

Because agriculture was never fully included in a GATT Round, many agricultural tariffs were not "bound" by the GATT process, so there were no hindrances for countries that wanted to increase their tariffs or other trade barriers. The European Community was allowed to adopt its Common Agricultural

Bound Tariffs

"Bound" tariffs are ones that have been included in the GATT process and essentially sanctioned at that rate. The tariff rate cannot increase from the bound level. Because the European Union tariffs had never been bound, they could increase without GATT objection.

Policy (CAP) and withdraw all its previous tariff bindings with concessions that were given on nonagricultural products (Hudec). Thus, the variable levy of the E.C. was GATT-legal.

Domestic subsidies should have been a violation of national treatment, because domestic producers were receiving benefits in the country that were not allowed to foreign producers. This was never enforced through a GATT panel because there was not a pro-trade climate when many of these policies were instituted. Subsidies became so pervasive among the more developed countries that they were never challenged.

By the mid-1980s, the costs of agricultural programs in many countries were becoming burdensome and there was great pressure to lower government expenditures (particularly in the E.C. and the United States). The United States was also committed to liberalized agricultural trade, even though it followed many policies that artificially encouraged production and exports while discouraging imports. American farmers, politicians, and the American public were convinced that the United States would gain much if world agricultural trade was liberalized. Thus, the Uruguay Round, the eighth and final GATT Round, began in September 1986 with agriculture as one of the specific areas for negotiations.

URUGUAY ROUND—TIMETABLE OF NEGOTIATIONS

The Uruguay Round of the GATT got its name from the location of the initial meeting in Punta del Este, Uruguay, in September 1986, where member nations committed themselves to "greater liberalization of trade in agriculture." Other important areas covered by the Uruguay Round were intellectual property, services, trade-related investment measures, and the dispute settlement procedures. These latter areas have impacts on agriculture, but the heart of the Round had very specific impacts.

Despite the general commitment to liberalize agricultural trade, there were many differences among the member nations on the way to do this and the extent of liberalization. Three important groups developed relative to the initial stages of the negotiations: the United States, the European Community, and the Cairns Group. The latter group was made up of Australia, Argentina, Brazil, Canada, Chile, Colombia, Fiji, Hungary, Indonesia, Malaysia, New Zealand, the

Philippines, Thailand, and Uruguay. The United States and the Cairns Group supported more open trade, while the European Community was more cautious. Other important players, such as Japan and Korea, stayed mostly on the sidelines to watch the other groups, hoping that the negotiations would support their restrictive-trade stance.

The negotiations got off to a rocky start when the United States proposed that all trade-distorting policies in agriculture be eliminated over a ten-year period. This caught observers throughout the world by surprise, including many groups in the United States, and was dismissed by many as simply a polar negotiating stance, not a legitimate proposal for ultimate enactment, since the United States highly supported agriculture through trade-distorting means. The Cairns Group preferred a freeze on existing price supports, coupled with a phased reduction in their levels. The E.C. favored reductions in support after world prices were at higher levels. Producer subsidy equivalents (PSEs) were suggested as the best devices for negotiating reduced levels of internal support, despite their problems (see Chapter 5).

Instead of paving the way for an agreement, the three proposals were divisive, resulting in a collapse of the GATT talks in December 1988. The talks were restarted in April 1989, but there was still much work to be done and the conflict between the United States and the E.C. was heated. The disagreements in agriculture threatened to jeopardize the entire Uruguay Round.

A U.S. proposal for an agreement emphasizing three areas was submitted in December 1989. Those areas were market access, export subsidies, and internal support. These are the same three areas of ultimate agreement for the Round with respect to agriculture. A more detailed plan with specific targets in those areas was submitted by the United States in December 1990, and the Cairns Group submitted another specific plan based on the U.S. proposal. Both plans called for the elimination of export subsidies and the conversion of all nontariff barriers to tariffs. The E.C. was not happy with those plans.

The director-general of the GATT, Arthur Dunkel, submitted a draft Final Act for the Uruguay Round in December 1991 that included all negotiating areas, including agriculture, intellectual property, and services. Dunkel tried to strike a compromise among the various agricultural plans, since agriculture was the major sticking point for the entire Round. This draft Final Act was approved by the GATT members as the basis for concluding the Uruguay Round, but the final negotiations were halted by the E.C.'s demand to weaken the agricultural provisions—effectively blocking the entire Round because of agricultural conflicts. But the United States and the European Community reached a compromise on export subsidies and internal support in November 1992 (the Blair House Agreement), leading to the final agreement in December 1993.

URUGUAY ROUND—TERMS OF THE AGREEMENT

The three areas included in the December 1989 proposal—market access, export subsidies, and internal support—served as the main elements of the agreement for agriculture. Despite being signed in 1993 and beginning in 1995, the base years for liberalization in these three areas were in the 1980s (1986–1988 for market access and internal support; 1986–1990 for export subsidies). This allowed the higher prices in world agricultural markets that occurred throughout the 1990s to count toward the Uruguay Round commitments (thus reducing the needed liberalization).

All countries are committed to transform their trade barriers to tariffs immediately, with a few exceptions. This tariffication process bound the tariffs for all products and established a ceiling on future tariff rates. All tariffs will be reduced by a minimum of 15 percent over a six-year period (a ten-year period for less developed countries, or LDCs) and by an average of 36 percent over all commodities (24 percent for LDCs). The average is calculated as a simple average of tariffs over commodities, so it is not weighted by the commodity's importance in trade. Thus, a country could choose to reduce high tariffs on products that are not heavily imported, so the agreement's minimum 15 percent reduction is important.

For products where imports are less than 5 percent of domestic consumption, there is a 3 percent minimum access opportunity in the first year of the agreement where imports are subjected to a preferential duty rate. This minimum access opportunity amount increases to 5 percent by the end of the agreement. The provision essentially established a tariff-rate quota for products where imports have historically been low.

The agreement calls for no new export subsidies for agricultural products and for cuts in current export subsidies. Over six years (ten years for LDCs), beginning in 1995, members are to cut the quantity of products with export subsidies by 21 percent and the total budgetary outlays for export subsidies by 36 percent. Because world prices rose markedly between the late 1980s and mid-1990s (and the level of subsidy needed for exportation has fallen), the quantity adjustments may be more constraining than the budgetary reductions. Genuine food aid is not counted as an export subsidy. Countries also agreed not to retaliate against existing export subsidies (through countervailing duties), the so-called peace clause.

The commitments for reducing internal support for agriculture were more sensitive because of the division over which policies count as internal support. All policies were divided into ones that distort trade (called Amber policies) and those that do not (Green policies). Only Amber policies are targeted for reductions.

Green policies include costs of general services (research, extension, inspection, marketing, and promotion), public stockholding, decoupled income support, crop insurance and income safety nets, and conservation programs. These programs do not encourage production of particular products and are, therefore, classified as non-trade-distorting. Amber policies include price supports, marketing loans, acreage payments, livestock payments based on animal numbers, input subsidies, and some subsidized loans. Exempt from the Amber category are direct payments to producers that are linked to production-limiting policies. These exemptions include U.S. deficiency payments for crops and the E.C. compensatory payments system.

The agreement calls for countries to reduce their aggregate measure of support (AMS) by 20 percent over six years (13 percent over ten years for LDCs). The AMS has three components: the implied tariff (equal to domestic price minus world price multiplied by the quantity of domestic production), direct payments to producers (measured by government expenditures), and other government outlays (measured by expenditures). Any producer assessments to pay for this AMS are subtracted.

All waivers and special exemptions for agriculture are removed by the agreement, and there are special provisions on sanitary and phytosanitary (SPS) regulations (rules that protect human, animal, or plant life and health from biological and chemical risks). Under the agreement, the SPS regulations must be based on science. Countries are encouraged to use international standards but are allowed to have their own, stricter regulations if they have scientific evidence and a risk analysis to back them up.

URUGUAY ROUND—ESTIMATED EFFECTS OF THE AGREEMENT

The U.S. Department of Agriculture's Economic Research Service (USDA-ERS) spent a great deal of time developing models to project the impact of the Uruguay Round agreement on world trade and incomes. Most of their work focused on what will happen to the United States, but they also estimated what will happen with other countries, since U.S. agriculture is very export-oriented. They made estimates for the year 2000, for shorter-term impacts, and the year 2005, for longer-term impacts. The longer-term impacts are generally larger than the shorter-term impacts because decision makers have more time to adjust to price changes. All estimates have ranges that are constructed in a manner similar to confidence intervals to reflect the uncertain aspects of the estimation process. The estimates are *ceteris paribus,* meaning that they concentrate only on the impacts of the agreement, everything else equal.

The USDA estimates that world income will increase by $5 trillion from 1995 to 2004 because of the new agreement and that demand for income-sensitive food products will grow markedly (ERS). This is brought about by more efficient use of world resources, since relatively high-cost producers will lower production in favor of lower-cost producers. Consumers will also increase purchases because world prices for most products will be lower.

TABLE 6.1 Range of Estimated World Effects for the Uruguay Round Liberalizations by the Year 2000

	Wheat	Corn	Soybeans	Dairy	Beef	Pork	Poultry
Output	3–7%	0%	1–2%	0%	*(2)–0%	1–3%	(3)–1%
Exports	7–11%	5–10%	3%	6–24%	7–11%	10–15%	9–25%
Imports					11–15%	(5–15)%	
Price	3–6%	0–5%	0–3%	0–2%	2–4%	0–3%	1–4%
World trade	(2)–0%	1–2%	0–1%	(1)–1%	0–4%	1–5%	2–5%

*Numbers in parentheses are negative, that is, (5)% is −5%.

Source: Economic Research Service.

TABLE 6.2 Range of Estimated World Effects for the Uruguay Round Liberalizations by the Year 2005

	Wheat	Corn	Soybeans	Dairy	Beef	Pork	Poultry
Output	5–7%	2–3%	2–3%	0%	*(1)–1%	1–3%	(1)–1%
Exports	9–12%	8–12%	2–3%	6–24%	10–14%	10–15%	25–36%
Imports					6–10%	(5–15)%	
Price	8–12%	6–9%	5–9%	0–2%	5–7%	3–5%	1–3%
World trade	0–2%	5–7%	1–3%	1–3%	4–11%	5–15%	3–7%

*Numbers in parentheses are negative, that is, (5)% is −5%.

Source: Economic Research Service.

Their projections for major U.S. agricultural products are presented in Table 6.1 for the year 2000 and Table 6.2 for the year 2005. Total U.S. agricultural exports are expected to increase by $1.6 to $4.7 billion by 2000 and $4.7 to $8.7 billion by 2005. Grains and livestock account for 75 percent of the increases. Tables 6.1 and 6.2 show that the livestock sectors are expected to enjoy larger percentage increases, especially poultry.

Cash receipts by U.S. farmers are expected to increase by $4.0 to $5.4 billion in 2000 and $9.8 to $11.6 billion in 2005. Farm income is expected to increase by $1.1 to $1.3 billion in 2000 and $1.9 to $2.5 billion in 2005. These numbers reflect expected higher prices for all major commodities and increased production levels for all commodities, except possibly beef and poultry.

The U.S. government will have lower tariff revenues because of the Uruguay Round agreement. The USDA's Economic Research Service (ERS) estimates that tariff revenue will fall by $275 million in 2000. If the United States had kept the deficiency and set-aside payments of the old farm programs, government spending in agriculture would have fallen by $300 to $700 million in 2000 and $1.7 to $2.5 billion in 2005. The Federal Agriculture Improvements and Reform (FAIR) Act of 1996 changed this, but the GATT agreement (with its expected higher producer incomes) made it easier for the FAIR Act to pass the U.S. Congress.

The major countries that made changes as a result of the Uruguay Round were the United States, the European Union, Japan, and Korea. Other countries made changes, but most are not expected to have major impacts on commodity trade. The U.S.–E.U. agreements on reducing export subsidies were key elements for grains, especially wheat and wheat flour. Subsidies for grains had heavily distorted world grain markets and made it difficult for nonsubsidizing nations to compete. Note, though, that the ERS expects only small percentage increases in grain trade as a result of the agreement through 2005.

World meat trade will increase more in percentage terms than grains. Beef trade will be impacted by lower E.U. subsidies, a lower Japanese tariff (38.5 percent rather than 50 percent), greater access to the Korean market (through negotiated minimum quantities and ultimately a 41 percent tariff instead of quotas in 2001), and the U.S. quota changing to a 31 percent tariff. These changes are expected to bring double-digit increases in U.S. beef exports by the year 2005. World pork and poultry trade increases as a result of liberalization, and U.S. exports are stimulated because of lower E.U. subsidies and the United States' opportunity to capitalize on greater access in many countries.

World dairy trade will be affected by lower E.U. and U.S. subsidies, and the United States will replace section 22 quotas with tariffs. The U.S. tariffs will be reduced by 15 percent over the agreement's life. Dairy exporting countries will have greater access to the Japanese market.

Japan and Korea, two of the most notoriously closed markets, agreed to significant liberalization across the board. All Japanese nontariff barriers will be converted to a tariff except for rice, which has a minimum access agreement. The Japanese must also follow policies that will constrain rice production and are not allowed to use export subsidies to dispose of rice surpluses. Korean nontariff barriers will convert to tariffs more slowly, and those tariffs will fall by 24 percent from 1995 to 2004. Korea also has a grace period on tariffication of rice quotas. It must follow policies to constrain rice production and cannot use export subsidies to dispose of rice surpluses.

THE WORLD TRADE ORGANIZATION AND THE FUTURE OF TRADE NEGOTIATIONS

The new World Trade Organization (WTO) supersedes the GATT because it is a more encompassing organization. Its main goals are to uphold the GATT and to provide a dispute settlement system that has more enforcement potential. The most important principles of the WTO are identical with those of the GATT: national treatment, most favored nation (MFN) treatment, and transparency. Regional agreements and special concessions for LDCs are the only exceptions to MFN.

The dispute settlement procedure is more formalized, making it stronger than the often confusing, ad hoc manner of settling disputes under the GATT (Hudec). Once a complaint has been lodged with the WTO, the parties are to enter consultations within thirty days. If there is no resolution of the conflict within sixty days, the complaining party can ask that a panel be assembled. If the two sides cannot agree

National Sovereignty and the WTO

A key question relative to the future of the WTO is, How willing are the large nations to allow WTO rules and panels to impinge on their national sovereignty? The WTO is very new, and the entry of the United States and other countries as member nations was controversial. Some people in the United States felt there was no reason why the largest and most powerful country in the world should have one vote on issues associated with world trade. Further, the United States might be required to abide by the ruling of a three-person panel concerning some policy enacted by the U.S. Congress. For example, Congress is considering import quotas on steel, which are clearly illegal according to the WTO.

The European Union has shown through its actions in the beef-hormone dispute that it doesn't like abiding by WTO rulings and is doing everything possible to avoid changing its policies. There are many such examples where national interests collide with the WTO. Yet if national interests are allowed to prevail, the WTO will collapse. One cannot allow powerful nations to escape WTO rulings and expect less powerful nations to be a part of the organization. Further, if national interests prevail in one case, the floodgates will be open for protracted future cases. There is no question that all WTO member nations must be willing to give up national sovereignty and interests if the WTO is to be an effective organization in world trade.

---

on membership of the panel, the director general of the WTO will appoint the panel. The panel will hear the case, and the final resolution of the issue must come within fourteen months of the original filing. The losing party must compensate the other party or withdraw the concessions covered by the complaint.

This new dispute mechanism makes it easier to resolve issues because there is a clear timetable for the settlement. There are also clear winners and losers. However, there is still no way that the winning party can guarantee that the losing party will ever follow through with compensation or change the offending policies. If there is no redress, however, the winning party can place sanctions on the offending country.

It is believed that the culture of dispute settlement under the WTO will be conducive to greater international cooperation. When nations consider joining the WTO, it is made clear that there will be more pressure to follow the outcomes from the dispute settlement procedures and that nations are expected to sacrifice some degree of national sovereignty to obtain the benefits of membership to this important international organization.

Future issues that will be important to the WTO include integrating the environment into trade negotiations (as discussed in Chapter 9), setting scientifically based technical standards, and settling rules for investment flows among countries (both direct investment and portfolio investment).

TWO OTHER IMPORTANT INTERNATIONAL AGENCIES

There are two other important international agencies that have a large impact on world trade, macroeconomic conditions, and development: The International Monetary Fund (IMF) and the International Bank for Reconstruction and Development (World Bank). Both of these organizations are officially under the UN, but they are truly independent administratively and financially. Both the IMF and the World Bank were created soon after World War II.

International Monetary Fund

The concept of and need for the IMF grew out of the global depression of the 1930s, which was partially caused by steep tariff increases and currency devaluations aimed at stemming balance of payments problems. Countries tried to reduce their problems by devaluing their currencies to increase their exports and reduce their imports. The world trading system collapsed after a series of devaluations caused tremendous instability in the world market.

The IMF was created to provide a pool of funds that countries can use to get out of temporary balance of payments problems. The IMF's goals are to help countries manage their payments balances efficiently and to help countries institute policy reforms that will help with a long-run resolution of these problems. If IMF loans are used properly, they will keep countries from devaluing their currencies to overcome short-term currency shortages. In order to meet these goals the IMF documents and monitors macroeconomic policies on all of its member countries.

The IMF conducts "missions" or studies for many LDCs with outstanding loans. The outcome of the mission is a set of recommended changes in policies and institutions that will assist in moving the country toward macroeconomic stability. Most IMF loans are made on the condition that certain policy reforms will be enacted (so-called structural adjustment programs). These adjustments might include cutting subsidies, selling state-owned enterprises, cutting the budget deficit, devaluing the currency, or other changes that are difficult for the country to implement. The terms of these policy changes are part of the loan negotiation process. If the policy changes are not enacted, the country will not be allowed to draw further on its line of credit or will not be approved for new loans. Because the IMF now deals in longer-term debts, it is a more important development financier than in its original charter.

Countries must sign rules of conduct and articles of agreement to become IMF members. They must allow free exchange of their currency, and there must be a relatively open macroeconomic policy system that is easily tracked by the IMF.

The IMF receives funds from its member countries, and the amount of money available for loans to a particular country is proportional to its subscription rate (or membership dues). The subscription rate for each country is based on international economic importance, and voting is based on contributions. The Fund is governed by a board of governors representing its members.

In recent years the IMF has had huge demands on its financial resources. It increasingly finds itself teaming with countries like the United States and Japan to meet the financial needs of many countries. The IMF, though, has a trained team of macroeconomic experts with international expertise in handling balance of payments crises. Its involvement in these difficulties brings an international perspective to the crisis that helps overcome the politics that often hamper unilateral programs.

International Bank for Reconstruction and Development (World Bank)

The World Bank's chief role is financing long-term development projects in LDCs. The Bank uses the funds it receives from its member nations, along with funds it borrows from international credit markets, to finance projects that would not normally be handled by the private sector. In the early years, many of the World Bank's projects involved water system development; road, bridge, and dam construction; agriculture; and other basic system needs. The Bank still funds many such projects, but now its role is more diverse given the wide range of needs throughout the developing world.

A World Bank project is developed in collaboration with the host government and is ultimately approved by the Bank's board of directors. Project loans from the World Bank go to host governments at concessional rates. It is normally the government's responsibility to manage the project. Over the years the Bank has increasingly been involved in structural adjustment loans that are contingent upon policy reforms in the country. Often the IMF and World Bank work in concert to determine the financial needs for structural reforms, and the Bank will come through with the needed long-term financing. Yet, there are times when the individual roles of the IMF and World Bank are clouded, especially when the two institutions do not agree on needed reforms or the timing of such reforms.

There are also regional development banks that work in a manner similar to the World Bank. These include the African Development Bank, the Asian Development Bank, the InterAmerican Development Bank, and the European Bank for Reconstruction and Development.

SUMMARY

1. The main guiding principles of the GATT are reciprocity, nondiscrimination, transparency, national treatment, and compensation.
2. GATT negotiations are normally handled through Rounds where specific sectors are targeted for trade barrier reductions.
3. The Uruguay Round began in 1986, and agriculture was its major focus. Other important areas covered by the Round were intellectual property, services, trade-related investment measures, and the dispute settlement process.
4. Initially in the Uruguay Round, there were three groups with proposals for agricultural trade liberalization: the United States, the European Union, and the Cairns Group. The talks collapsed in December 1988, but after much negotiation between the United States and the E.U. (and a major reform of the E.U.'s Common Agricultural Policy), a final agreement was reached in December 1993.

5. The Uruguay Round agreement for agriculture involved three main components. All trade barriers were transformed into tariffs, and these tariffs will be reduced by a minimum of 15 percent (an average of 36 percent) over six years. Countries must reduce their aggregate measure of support by 20 percent over six years. Countries cannot introduce new export subsidies, and they must reduce existing export subsidy programs by 21 percent in quantity and 36 percent in value.

6. The World Trade Organization was formed as a more encompassing organization to supersede the GATT. The WTO has a stronger dispute settlement process and more means to enforce rules.

7. Two other important international economic organizations are the IMF, which loans money to governments for stemming balance of payments problems, and the World Bank, which finances long-term development projects in less developed countries.

QUESTIONS

1. Why were the Uruguay Round GATT negotiations so important for agriculture? Why were they so difficult?
2. Are you in favor of U.S. membership in the WTO, or did the United States sacrifice too much sovereignty? Explain.
3. Would you be in favor of the United States' initial stance in the Uruguay Round—to eliminate all government support to agriculture? Why?
4. What agricultural product trade has been most influenced by liberalization resulting from the Uruguay Round?
5. What are the key agricultural trade issues for the next GATT Round?

REFERENCES

Hudec, Robert. "Does the Agreement on Agriculture Work? Agricultural Disputes after the Uruguay Round." International Agricultural Trade Research Consortium Working Paper #98-2, April 1998.

Economic Research Service (ERS), U.S. Department of Agriculture. "Effects of the Uruguay Round Agreement on U.S. Agricultural Commodities." GATT-1. Washington, DC: Government Printing Office, March 1994.

———. "Agriculture in the WTO." Situation and Outlook Series. WRS-98-4. Washington, DC: Government Printing Office, December 1998.

Chapter 7

Preferential Trade Agreements

Despite all the progress on trade liberalization that has been made through the GATT since 1946, there have been numerous efforts to form alliances among countries for free trade. These alliances, which allow liberalized trade among the members, are called *preferential trade agreements* (PTAs), and they come in two general forms: the free trade area (FTA) and the customs union (CU). The best-known FTA is the North American Free Trade Agreement (NAFTA) among Canada, Mexico, and the United States; while the best-known CU is the European Union (E.U.) among most countries of Western Europe.

This chapter covers the theory of preferential trade agreements and the situations when member countries gain and lose. The key factor determining whether countries gain or lose is what happens to tariff levels. If they fall, the country will likely gain; if they don't, the country might lose tariff revenue and import from relatively inefficient member countries, reducing welfare of the member nations. The best-known FTA (NAFTA) and CU (the E.U.) are covered in some detail to provide an idea of what it is like to form and operate under PTAs.

INTRODUCTION TO PTAS

Customs unions have no trade barriers among members, and the members also have a common external tariff (for nonmembers). Often a CU will have free movement of production factors (labor, investment capital, etc.), harmonized standards, and even a common currency (in the case of the E.U.). Free trade areas allow the member countries to keep their own trade barriers with nonmembers, but there are rules of origin that keep nonmember goods from being reexported between FTA members.

Preferential trading agreements violate the GATT/WTO principle of most favored nations: once a trade barrier is lowered for one GATT-member country, it should be lowered for all member countries. However, Article 24 of the GATT allows an exemption for PTAs as long as they involve 100 percent lowering of tariffs among their members (i.e., when tariffs are eliminated). When trade barriers have not been completely eliminated (which is the case with NAFTA), the GATT has viewed them as coming close enough to get approval under Article 24. Often PTAs involve a transition period where tariffs and other trade barriers are slowly scaled to zero. A PTA will not receive a GATT exemption if the agreement raises tariffs for non-PTA members.

The United States was initially against Article 24 because it thought that PTAs (and regional trade agreements or regionalism) would undermine multilateral negotiations through the GATT. As will be seen later in this chapter, there is a theoretical dispute concerning whether PTAs increase global welfare. There are trade creation effects, which are positive, and trade diversion effects, which are negative. The United States was quite vocal in its opposition to PTAs because there are only trade creation effects from multilateral trade liberalization through the GATT.

Yet the United States' attitude changed in November 1982, when a new GATT Round was not launched, against the strong wishes of the United States. The U.S. immediately began negotiating bilateral agreements with Canada and Israel the following year, arguing that bilateral agreements and PTAs would move countries into multilateral negotiations.[1] Since 1982, there has been a proliferation of PTAs throughout the world and, during the early 1990s, there was much talk of creating many others.

This chapter will cover the economic aspects of PTAs and go over conditions when the world benefits from their creation and when the world loses from their creation. Later in the chapter the two largest PTAs, NAFTA and the E.U., will be discussed to highlight the negotiation process, rules concerning their operation, and changes they have brought to the rest of the world.

WELFARE ASPECTS OF PTAS

The main distinguishing factor between PTAs and pure trade liberalization is that there is a tendency for exports to be diverted from low-cost suppliers to higher-cost suppliers as a result of trade preferences. Table 7.1 provides a simple example of the negative aspects of trade diversion. The trade creation and diversion examples include a common external tariff for the countries for simplicity, so the PTA is a customs union. Assume that countries A and B enter into a CU but country C is excluded. Before the agreement, each country imports wheat from country C, the low-cost wheat producer (as shown in column 2 of Table 7.1). The price for a bushel of wheat is $6, $4, and $3, respectively, in countries A, B, and C because of the $3 tariff ($t$) for country A and the $1 tariff for country B. This analysis assumes that transportation costs are zero and there is perfect competition, so price differentials among countries reflect tariffs.

Trade diversion would occur in the wheat example if there is a set of policies that encourage country A to import wheat from country B. The reason is that before

---

[1]In 1988, the U.S. Omnibus Trade Bill included Super 301 provisions, which are another way to circumvent the GATT dispute settlement procedures for bilateral disputes.

TABLE 7.1 Examples of Trade Creation and Trade Diversion*

Country	Before CU	After CU (*TD*)	After CU (*TC*)
A	$6 ($t = 3$)	$6 ($t=3$)	$4 ($t=1$)
B	$4 ($t = 1$)	$6 ($t=3$)	$4 ($t=1$)
C	$3	$3	$3
Trade pattern	C to A	B to A	C to A
	C to B		C to B

*t is tariff.
TD is trade diversion case.
TC is trade creation case.

the CU, country A was importing from the low-cost producer. Importing from country B would be encouraging higher-cost producers to increase wheat production and would therefore be inefficient (and decrease world welfare). Column 3 shows a situation where the external tariff for the CU is $3 per bushel, the same tariff that prevailed in country A before the CU. This tariff would result in increased production in country B because the internal price in B increases by $2 per bushel. Some of that increased production in B would be shipped to A (and crowd out imports from country C), so trade would be diverted from exports of country C to country B. Trade diversion occurs because the cost for the added production in country B is greater than the cost for the production in country C, which was "crowded out."

Trade creation would occur in the wheat example if production in country B is not encouraged, as shown in column 4. In this case, the common external tariff is set at $1 per bushel, identical with the initial tariff in country B. There is no increase in B's production, but there are trade creation benefits because of the lower price in country A and the increased exports from country C to country A. Wheat continues to move from country C to country B, as it did before the customs union.

Examples of trade diversion and trade creation are shown graphically in Figures 7.1 through 7.3. Figure 7.1 shows the trade diversion effects of a customs union between countries A and B. ED_a is the import demand curve for wheat in country A. ES_b and ES_c are the export supply curves for countries B and C, respectively, and both are assumed to be perfectly elastic (so country A is a small country). Without the customs union, if the tariff for country A is t, it imports Q_1 units of wheat from country C and country A receives tariff revenue of area $abfg$. There is no trade between country A and country B.

When the customs union is formed, there is free trade between countries A and B, so the price in country A falls to P_b (as long as country B's excess supply curve is perfectly elastic) and country A imports exclusively from country B as long as the external tariff is greater than $P_b - P_c$. Country A gains surplus area $abce$ because its internal price has fallen to P_b, but it loses area $abfg$ because it no longer collects tariff revenues.[2] One must compare the bde triangle with the $cdfg$ rectangle to determine whether country A gains from the CU for the wheat market.

---

[2]If country A continues to import from country C, then it will receive some of the tariff revenue.

FIGURE 7.1
Results of a customs union with trade
diversion.

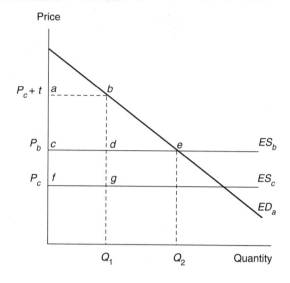

FIGURE 7.2
Gains for country A from the customs
union with country B—trade diversion.

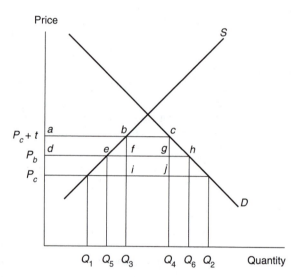

Figure 7.2 pinpoints the gains from the customs union accruing to country A
when there is trade diversion. In the figure, the original internal price in country A
is $P_c + t$ and all imports are from country C. Imports are $Q_4 - Q_3$, and tariff revenue
for country A is area $bcij$. When a CU is formed with country B, the price falls to P_b
and imports increase to $Q_6 - Q_5$, but tariff revenue is zero because all wheat comes
from country B. The increase in consumer surplus is area $acdh$, the reduction in producer surplus is area $abde$, and the decrease in government revenue is area $bcij$. The
net gain in the wheat market resulting from the CU is $bef + cgh - fgij$, which is ambiguous. If the price difference between country B and country C is small, there
should be net gains to country A. Note that imports with free trade are $Q_2 - Q_1$.

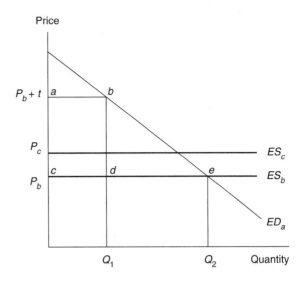

FIGURE 7.3
Results of a customs union with trade creation.

There is no question that country A would gain if it had a CU with country C, because it would be obtaining the benefits of full trade liberalization, but gains may be negative for the CU with country B. The key element that causes trade diversion is that the discriminatory trade policies encourage country A to import from higher-cost producers in country B, diverting imports from the low-cost producers in country C. This crowding out of production in country C with production in country B also makes the world worse off.

Figure 7.3 covers the case when the customs union between country A and country B is trade-creating. In this case, country B is the low-cost producer and the gains from trade are the familiar trapezoid *abce.* The price in country C is above the price in country B, so there is never an incentive for country A to import from country C. Country A clearly gains from the CU because the surplus gain more than compensates for the loss in tariff revenue of area *abcd.*

These examples show the basic idea behind trade creation and trade diversion, yet the assumption that country B's export supply curve is perfectly elastic is simplistic. In the next case, assume that country B has an upward-sloping export supply curve, while country C has a perfectly elastic export supply curve.[3] Trade diversion with an upward-sloped export supply curve for country B is shown in Figure 7.4.

In Figure 7.4, the equilibrium before formation of the customs union has imports from countries B and C facing the tariff, t. The internal price in country A is $P_c + t$, resulting in imports of Q_1 and $Q_3 - Q_1$, respectively, from countries B and C. Country A's gains from trade are area *abd* plus area *bdeh* (the tariff revenue, t times Q_3).

---

[3]In this analysis, one can assume that country C represents the rest of the world. The fact that its export supply function is perfectly elastic represents the assumption that country A and country B are small countries.

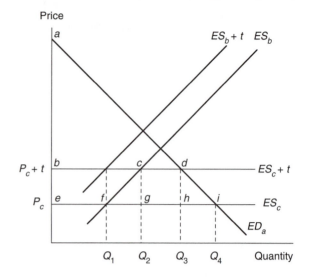

FIGURE 7.4
Trade diversion with rising costs for country B.

When the CU is formed and imports from country B are allowed into country A duty-free, there is no change in country A's internal price as long as the tariff remains the same, so the surplus area *abd* does not change. However, there is a reduction in the amount of tariff revenue because imports from country B are not taxed. Because of duty-free access, exporters from country B are able to supply wheat to country A at a lower price, increasing their exports to Q_2 and crowding out exports from country C by $Q_2 - Q_1$. The net welfare effect on country A is that welfare falls by area *bceg*, the fall in tariff revenue.

Country B is the winner from the formation of the customs union for wheat because it exports more to country A. Its gain is area *bcef* because its exports grow from Q_1 to Q_2. Yet when the gains to country B are summed with the loss of country A, net welfare for the two countries falls by area *cfg*—the deadweight loss associated with higher-cost producers in country B crowding out producers in country C. The best course for country A is complete liberalization with all countries, which would provide a net increase in welfare of area *dhi* because the internal price in country A would then reflect world production costs of P_c. Country A would increase imports to Q_4, importing Q_1 from country B and $Q_4 - Q_1$ from country C.

Figure 7.4 was drawn so that country A continues to import from country C with the formation of the customs union. This is not necessarily the case. If ES_b intersected ED_a at a point beyond Q_3 (as shown in Figure 7.5), country A would import exclusively from country B. In this case, country A would begin to experience a lower internal price, consistent with the price in country B (P_b). Country A would have a surplus gain of area *abcd* from the lower internal price, but would lose all of its tariff revenue (area *abef*) because no imports would arrive from country C. There would be deadweight losses because production shifts from country C to country B, but the more the intersection of the ES_b and ED_a curves lies to the right of Q_3, the

FIGURE 7.5
Trade diversion with more efficient
producers in country B.

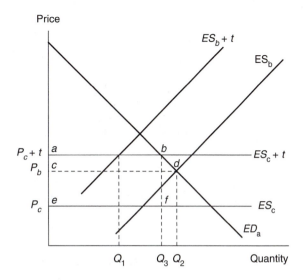

lower the net losses to country A. These lower net losses are because country B's producers are more efficient than in Figure 7.4, so the losses in bringing them into production to supply country A are lower.

It should be noted that this analysis is for a product that is imported by country A and exported by country B. In this case, country B gains and there is a possibility (if trade diversion is large) that country A will lose. Yet there are other commodities where country A will export to country B and country A will therefore have clear net gains from the customs union. It is these gains in other commodities that encourage country A to enter into the CU.

The economic analysis of welfare gains from PTAs has focused on a discussion of trade diversion because this concept is new for this chapter. PTAs involve many possibilities concerning the ultimate tariff levels that pertain to countries involved in the agreement. The levels of these tariffs for nonmembers are what leads to trade creation versus trade diversion. If a country enters a PTA and simultaneously lowers its external tariff on nonmembers (or if countries entering a CU agree to establish the common external tariff at the lowest level among member countries), the possibility for trade diversion effects is lessened.

One final comment on how preferential trade agreements distort behavior involves foreign direct investment. PTAs will not only divert trade but also change investment patterns among countries. Many companies have chosen to establish facilities in Europe to gain access to lower trade barriers among E.U.-member countries. Asian and European companies have considered investing in Canada and Mexico to gain access to NAFTA treatment and free trade among the three member countries.

These changes in foreign direct investments can generate misleading statements and understandings about the effects of PTAs. The U.S. textile industry has been characterized by sustained contraction for many years because of firms moving abroad to take advantage of lower costs. In recent years, some of these firms have left for Mexico, and frequently observers blame their exits on NAFTA. This observation is partially true. The firms decided to locate in Mexico because the tariffs they face when exporting back to the United States might be lower in Mexico. However, their decision to exit the United States was not due to NAFTA but to lower cost structures for textile manufacturing outside the United States. The fact that they chose to locate in Mexico rather than Costa Rica, Honduras, Indonesia, or elsewhere is due to NAFTA. This is another example of trade diversion, since textiles will flow from Mexico to the United States, rather than from other countries that have lower cost structures than Mexico.

NORTH AMERICAN FREE TRADE AGREEMENT

An example of a free trade area is the North American Free Trade Agreement (NAFTA), which began in 1986 with negotiations between the United States and Canada. The discussions leading to the final agreement, which was called the Canada–U.S. Trade Agreement (CUSTA) at the time, were relatively painless and moved rather quickly. That agreement went into effect on January 1, 1989, where all bilateral tariffs were phased out over a five- or ten-year period, depending on the product (though temporary "snapback" provisions existed to prevent import surges). In addition, a number of bilateral working groups began to harmonize rules and regulations imposed to protect animal, plant, and human health. Now trade between the United States and Canada has few hindrances, and tariffs are zero, though there are still institutions (such as the Canadian Wheat Board) and policies (such as for U.S. sugar) that impede trade.

Mexico's movement toward free trade also began in 1986, when it joined the GATT. This was a signal that Mexico was willing to open its economy to the rest of the world and suffer the adjustments necessary to obtain the benefits of more liberalized trade. Mexico was forced into this decision because it had failed so miserably with its past policies of protection. In 1986, it had a foreign debt of $99 billion, second only to Brazil's $111 billion, and debt service payments were 79 percent of its export revenues (International Monetary Fund). Gross national product had fallen by 10 percent in 1986, and inflation was running at an 86 percent annual rate. Over 92 percent of Mexico's production was protected through import licenses, which were required before products could enter the country (Congressional Budget Office, CBO). The average tariff rate (weighted by each product's proportion of trade) was 25 percent and some tariffs were as high as 100 percent. Mexico had a heavily protected economy that was performing very poorly.

Mexico underwent an economic metamorphosis between 1986 and 1990, unilaterally liberalizing to meet the GATT guidelines and surpassing them by becoming more open than the GATT requires. By 1990, economic growth was restored, import licenses protected only 19 percent of Mexico's imports (though most of them were for agricultural products), the average tariff was only 12.5 percent, and the maximum tariff was 20 percent (CBO). The government of Mexico was attempting to tell the world that it was ready to become a free-market economy with tremendous growth potential.

In August 1990, President Carlos Salinas de Gortari officially requested that the United States and Mexico enter into negotiations for a free trade agreement. Within five weeks, President George Bush formally notified Congress that he intended to begin negotiations with Mexico. Canada decided to enter the process in February 1991, since it already had a free trade agreement with the United States. Finally, Congress gave the President "fast track" negotiating authority in May 1991 (allowing the executive branch to negotiate the details of the agreement and Congress to vote on the overall agreement), and negotiations began in June.

Mexico's request for a free trade agreement reflected its desire to institute permanently the reforms that had already taken place and to continue the trade liberalization momentum for the Mexican economy. The country was particularly interested in attracting foreign investment, which it had openly discouraged for many years. Foreign investment was restricted to 49 percent of a project in 1986, but 100 percent foreign ownership was allowed on projects up to $100 million by 1991. The approval procedures were also streamlined with fewer "bureaucratic" hassles to overcome.

Free trade with Mexico was not a new concept for the United States, though many people did not know that when the NAFTA negotiations commenced. Many exports from Mexico received duty-free access because of the generalized system of preferences. However, trade between the United States and Mexico was even more affected by the maquiladora program. Started in 1965, it allowed "free trade" between northern Mexico and the United States. Firms locating in specific areas around the U.S./Mexico border, often operating twin plants on each side of the border, are allowed to ship intermediate and finished products across the border without duty on the value-added as long as the goods have not exited the free trade zone.

When the maquiladoras were set up, most of them were textile firms that sewed garments and performed other low-skill activities. Now many of them are electronics and transportation equipment firms that perform highly skilled activities. The number of maquiladora plants has grown by 20 percent annually since 1965, and they currently employ 600,000 people (*The Economist*). In 1991, they accounted for 37 percent of Mexico's exports and 23 percent of Mexico's imports. Naturally, they are extremely important for Mexico's bilateral trade with the United States, accounting for 46 percent of exports to the United States and 32 percent of imports from the United States (CBO). The maquiladora program allowed manufacturing companies to access the United States with finished goods and to increase the amount of processing performed in Mexico. The program, however, has had minimal impacts in the agricultural sector.

Many bulk agricultural imports into Mexico faced licensing requirements in 1991. The average tariff on U.S. agricultural exports to Mexico was only 11 percent (Runge), yet Mexico's producer subsidy equivalents (PSEs) for crops averaged 40 percent, while its PSEs for livestock averaged below zero (Nelson, Simone, and Valdes). Mexico's PSEs for corn, soybeans, and wheat were particularly high, discouraging imports from the United States.

The United States had lower overall barriers on agricultural imports from Mexico, but there were certain tariffs that varied by season to protect U.S. producers, particularly for fruits and vegetables. Often these seasonal tariffs were particularly high when Mexican produce was harvested, which was frustrating for its producers. The average U.S. tariff on Mexican agricultural goods was only 5 percent, though.

Basic Elements of the Agreement

Canada, Mexico and the United States completed negotiations and released a draft accord to create a free trade area in August 1992. The document was signed by President Bill Clinton in December 1992, but he said that he would not forward it to Congress for approval by the U.S. House of Representatives until there were side agreements on labor standards and the environment. These side agreements were negotiated in 1993. The NAFTA proposal passed Congress in November 1993 and was signed by President Clinton on December 8, 1993, becoming effective January 1, 1994.

As detailed in Chapter 9, the NAFTA process allowed a discussion and resolution of environmental issues that had plagued U.S.–Mexican affairs for many years. The maquiladora program had generated increased economic activity that quickly outgrew the capacity of the border areas to handle pollution. Factories dumped toxic waste into the rivers, spewed air pollution, and fouled all areas of the border. Uncontrolled industrial growth, undeveloped infrastructure to handle pollution, a lack of enforcement of environmental regulations, and the like, had turned the border into a pollution nightmare (Runge).

The 1983 La Paz agreement between the United States and Mexico on Cooperation for the Protection and Improvement of the Environment in the Border Area instituted many regulations that were not followed, including construction of waste treatment plants, planning for hazardous waste spills, and air emission programs (Runge). One requirement was that imported toxic chemicals used in maquiladora production be exported back to the country of origin as toxic waste. This lack of enforcement was very disconcerting to the U.S. environmental community, and these concerns have entered into their position on NAFTA issues.

Mexico unilaterally attacked some of its environmental problems through enactment of the General Law for Ecological Equilibrium and Environmental Protection in 1988, which is modeled on U.S. law. It requires companies to file plans and environmental impact assessments for any new construction and plant changes. A further agreement for the border areas resulted in the Integrated Environmental Plan for the Mexico–U.S. Border Region that was jointly adopted by the United States and Mexico in 1992. With these new environmental laws and agreements, there was increased enforcement: some factories have been closed, and others have been forced to reduce emissions by 70 percent (Runge). But there was still a question by many in the United States whether Mexican authorities were truly committed to pollution reduction and enforcement of Mexican laws.

During 1993, the United States and Mexico took time to discuss environmental issues associated with NAFTA and to develop a strategy to attack the problems. The result was a side agreement on the environment called the North American Agreement on Environmental Cooperation (NAAEC). The United States and Mexico also agreed to establish a Border Environmental Cooperation Commission.

The NAFTA addresses three components relative to the environment: pollution havens; human, animal, and plant health and safety; and international environmental agreements (Ballenger and Krissoff). With respect to pollution havens, NAFTA renounces the relaxation of environmental standards and establishes compulsory consultations among the member countries when relaxation occurs. Sanitary and phytosanitary provisions encourage members to use international standards, but countries can enact more stringent standards if they are scientifically backed. Countries should consider risk assessment, production methods and practices, scientific evidence, ecological and environmental conditions, and economic conditions in their choice of standards. The NAFTA protects national commitments to international environmental agreements to guard against challenges that individual countries might make. It preserves the status quo when an international environmental agreement is involved.

The NAAEC attempts to strike a balance between environmental aims and national sovereignty in relations among NAFTA members. It established the Commission on Environmental Cooperation (CEC), which is charged with examining environmental issues in the region, including production processes and methods (PPMs). The NAAEC commits the countries to enforcing laws, including training inspectors, monitoring and reporting on compliance, and using sanctions when violations occur. There is a dispute settlement procedure in the NAAEC, but it is not clear how effective it will be.

The NAFTA includes a transitional phase of five to fifteen years (depending on the commodity) for trade barriers to reach zero. Most nontariff barriers (including licensing requirements) were converted to a tariff equivalent (i.e., tariffication) immediately so that progress toward free trade could be monitored more easily. Tariff-rate quotas (TRQs) were established for some commodities, which allowed a certain quantity of imports at one tariff rate (often duty-free) with higher import volumes facing a higher duty rate. The higher duty rate was usually the most-favored-nation (MFN) rate.

The U.S. Trade Commission estimated that almost 54 percent of Mexican exports to the United States were duty-free in January 1994, compared with 31 percent of U.S. exports to Mexico. The agreement called for 62 percent of Mexico's exports to be duty-free after five years, compared with 47 percent of U.S. exports to Mexico. Only 1 percent of U.S. and Mexican imports were to be included in the fifteen-year liberalization plan. These were the most contentious products in the negotiations, and many of them were agricultural goods (corn and beans on the Mexican side and orange juice on the U.S. side).

Both the United States and Mexico have continued unilateral agricultural policy changes that have brought their agricultural sectors into a market-oriented situation. The PROCAMPO program in Mexico, which was announced in 1993 and became fully effective in 1995, substituted direct payments to agricultural producers based on acreage for input and commodity subsidies. Before 1995, Mexican agricultural producers had substantial subsidies for credit, irrigation, and fertilizer. The U.S. agricultural policy system was revamped under the Federal Agriculture Improvements and Reform (FAIR) Act of 1996, as discussed earlier. Thus, both countries have moved toward agricultural policies that are less output-distorting.

The NAFTA and the unilateral policy changes in Mexico and the United States were supposed to provide a tremendous boost to trade between the two countries. It was expected that U.S. agricultural exports to Mexico would increase markedly for many important commodities. However, this didn't happen because of the peso crisis in Mexico and the subsequent economic slowdown in the country.

Mexico's Peso Crisis

When Mexico opened its market to foreign capital, it was hopeful that it could not only attract foreign investors but also reverse the "capital flight" of previous years. The changes in investment policies, the commitment to increased economic liberalization, and the NAFTA deal helped woo foreign investors back into Mexico, and much of the flight capital returned.

Mexico was in a precarious financial situation, though, in 1994 because most (67 percent) of the capital inflow into the country was in portfolio investments that could easily be withdrawn. Mexico was running a current account deficit of 7.6 percent of GDP, and its foreign reserves had fallen from $25 billion to $6 billion in less than a year (*The Economist*). The short-term future looked even more bleak because the Mexican government had $29 billion in short-term tesobonos bonds that were indexed to the U.S. dollar (so the government suffered the exchange rate risk) and were due in 1995. President Carlos Salinas, who had championed much of the

Mexican liberalization, was leaving office on November 30, 1994, and his administration allowed these problems to carry forward to President Ernesto Zedilla's term of office.

With foreign exchange reserves at low levels, a huge need for money to repay the tesobonos bonds, and continued demand for spending by the government and consumers, it was clear that some action was needed. President Zedilla announced on December 20, 1994, that the Mexican peso would be devalued by 15 percent against the U.S. dollar. This announcement generated a nearly instantaneous $5 billion flight of capital out of Mexico. Investors had lost confidence and wanted out no matter what the cost. With low foreign exchange reserves, the Mexican government could only watch the downward spiral of its peso and stock market. The peso went from 3.5 pesos per dollar in December to 7.5 in early March. The stock market lost one-half of its value during the same period.

The United States arranged a $50 billion financing package to stabilize the situation in Mexico. The Mexican government allowed interest rates to climb (they reached over 80 percent for a short time), increased the value-added tax from 10 to 15 percent, and cut public spending by 10 percent (*The Economist*). All of these measures were used to cut the demand for borrowing by the government and consumers. The economic pain in Mexico was great during 1995 and early 1996, but the country has returned to positive economic growth.

The peso crisis had a huge impact on U.S. trade with Mexico. U.S. agricultural exports to Mexico, which had increased by 27 percent in 1994, fell by 22 percent in 1995. Mexican agricultural exports to the United States, which had increased by only 6 percent in 1994, increased by 32 percent in 1995. Both of these events in 1995 are consistent with the decreased value of the peso, which discouraged imports into Mexico and encouraged exports from Mexico. When the peso increased in value during 1996, trade with Mexico returned to a more normal course relative to 1994: U.S. agricultural exports to Mexico increased 54 percent in 1996, while Mexico's agricultural exports to the United States fell by 2 percent.

NAFTA Effects

Studies have attempted to isolate the trade and welfare effects of NAFTA. For trade involving Mexico, one challenge is to capture and eliminate the effects of the Mexican peso crisis. Nelson and his associates estimated the individual commodity effects of NAFTA for Canada, Mexico, and the United States. In aggregate, they estimated that U.S. agricultural exports are 3 percent and 7 percent higher, respectively, to Mexico and Canada because of NAFTA. They further estimated that U.S. agricultural imports are 3 percent and 5 percent higher, respectively, from Mexico and Canada as a result of NAFTA.

Some of the commodities experiencing the biggest gains resulting from NAFTA (each was over 15 percent higher) were U.S. exports of cattle, dairy products, apples, and pears to Mexico; U.S. imports of peanuts and live cattle from Mexico; U.S. exports of beef and processed tomatoes to Canada; and U.S. imports of beef from Canada. According to Nelson and his colleagues, some other products that experienced large changes in trade, such as U.S. imports of winter vegetables from Mexico, have been mostly influenced by weather patterns, the peso crisis, and technological shifts—phenomena outside NAFTA.

Tweeten, Sharples, and Evers-Smith estimated that NAFTA increased U.S. agricultural exports to Canada by $1.43 billion (or 25 percent) in 1995 and increased Canadian agricultural exports to the United States by $1.88 billion (or 34 percent). Their study also estimated that U.S. and Canadian agricultural producers lost $705 million and $776 million, respectively, because of NAFTA, but U.S. and Canadian consumers gained $961 million and $817 million, respectively, from NAFTA. These figures generate net gains to the United States and Canada of $256 million and $36 million, respectively, after reduced tariff revenues are considered.

Much of the increase in trade with Canada due to CUSTA has been in processed food products. Munirathinam, Reed, and Marchant estimated that free trade with Canada increased U.S. exports of many consumer-oriented products more than bulk or intermediate agricultural products. They estimated that U.S. exports of some consumer-oriented products increased from 45 to 88 percent as a result of NAFTA, whereas U.S. exports of some intermediate products and bulk products didn't increase at all, though most were in the 20 to 30 percent range for both types of goods. One would anticipate that CUSTA would have a larger impact on more processed food items because tariffs are generally higher for such products. Trade liberalization will have more of a price impact, and more processed food products likely have a higher price elasticity of demand.

EUROPEAN UNION

The customs union in Europe, which is now called the European Union (E.U.), was launched in March 1957 when the six members of the European Coal and Gas Community (France, West Germany, Italy, Belgium, the Netherlands, and Luxembourg) signed what is now known as the Treaty of Rome. That treaty established a European Economic Community (EEC). From the beginning, the EEC had a common agricultural and transportation policy, but there was also talk of much wider integration among the members.

Over the years there has been a slow but steady move toward not only enlarging the membership but also making the union stronger and more encompassing. There are now fifteen members (with a population of 370 million), a common currency (the euro), and a central bank (the European Central Bank).[4] Formal applications for membership have been filed by Bulgaria, Cyprus, the Czech

[4]The fifteen members of the European Union are Austria, Belgium, Britain, Denmark, Finland, France, Germany, Greece, Ireland, Italy, Luxembourg, the Netherlands, Portugal, Spain, and Sweden. Britain, Ireland, and Denmark joined in 1973; Greece joined in 1981; Spain and Portugal joined in 1986; and Austria, Finland, and Sweden joined in 1995.

Republic, Estonia, Hungary, Latvia, Lithuania, Poland, Romania, Slovakia, and Slovenia. Other nations are also interested in membership.

A huge step toward a truly unified market was the Single European Act of 1987, which came to be known as EC-92 because all of the adjustments were to be completed by 1992. The Single European Act legislated the formation of a true single market with harmonized product standards, no fiscal barriers to trade, no border checks, unified technical regulations, and so on. There was quite a bit of controversy among members concerning product definitions, safety standards, and other technical considerations, but the resulting harmonization has been great for European businesses attempting to sell in all E.U. markets.

The final piece to the economic puzzle of integration was a common currency. It was long known that currency fluctuations among the member countries were a hindrance to trade, but the move toward the euro took many years. It began with the Exchange Rate Mechanism (ERM) in March 1979, to which all countries belonged except Britain (which joined later). The ERM established exchange rate bands for currencies that were within 2.25 to 3 percent of a specified rate, and countries were pledged to keep their currencies within those values. Because of speculative attacks on some currencies in the early 1990s, the ERM bands were widened to 15 percent in 1993 until the euro was instituted.

The Maastricht Treaty on European union, which came from the Maastricht Summit in December 1991 and was signed in early 1992, provided the basis for monetary union. The treaty established five convergence criteria for countries to enter into the euro. These concerned government budget balance, public debt, inflation, interest rates, and exchange rate stability. The Maastricht Treaty was approved by all member countries, and eleven countries joined the European Monetary Union (EMU) system on January 1, 1999.[5] Euro notes will replace all national currencies in 2002.

The European Union has five important governmental bodies that oversee its operations: the Council of the European Union, European Parliament, European Commission, European Court of Justice, and European Central Bank. Their operations and functions have changed over time as the membership has changed and the idea of a European union has progressed. The European Parliament, in particular, has increased its power over the years.

The Council of the European Union is the supreme decision-making body and is made up of the heads of state from each member country. Each representative serves the legislative interest of his or her own country. The Council makes policies, coordinates policies among member states, and makes decisions related to treaties. Decisions are made unanimously, by majority or by qualified majority (about 70 percent), depending on the issue. Under the qualified majority, the larger countries have a larger share of the votes.[6] The presidency rotates among member state representatives each six months. There are also specialized councils made up

---

[5]Britain, Denmark, Greece, and Sweden did not join the EMU. Greece did not meet the convergence criteria, while the other countries chose not to participate.

[6]For instance, Germany, Britain, France, and Italy each have 10 votes out of the 87 total; Luxembourg has only 2.

of country representatives, such as the Council on Agriculture, which is composed of the various agriculture ministers.

The European Commission is the executive branch of the European Union. It proposes policy, drafts implementation measures, and supports legislation that is being considered by Parliament. The Commission also represents the E.U. in trade negotiation rounds. Each country has at least one Commission member; Germany, Britain, France, Italy, and Spain have two. They are appointed by member governments (subject to approval of Parliament) for four-year terms, but after appointment, the Commissioners are independent of the member governments and represent the entire E.U. The Commission carries on the day-to-day business of the E.U. There is a commissioner responsible for various portfolios (for instance, there is an agriculture commissioner).

The European Parliament is the legislative body, and each member represents people throughout the European Union. It has 626 members, and each country has a membership based on population (Germany has the most, 99; with Britain, France, and Italy at 87; and Luxembourg with 6). Members of the European Parliament are directly elected for five-year terms, and they clearly represent E.U.-wide interests. The Parliament has always voted on the E.U. budget, but the Maastricht Treaty gave the Parliament equal footing with the Council in making certain decisions. Depending on the decision, the Parliament will have a consultative role, a cooperative role, or a co-decision role.

Originally, these E.U. organizations would work in the following general way: the Commission would propose and the Council would decide, after consulting Parliament. For instance, the Commission on Agriculture would study target price rates and make a recommendation to the Council, which would decide upon the target price. However, Parliament's role has been widened over time so that its power has increased. For instance, now all Commissioners must be approved by Parliament.

The European Court of Justice interprets the Treaty of Rome and other E.U. agreements or actions in legal terms. It is composed of thirteen judges, each serving a six-year term. The Court of Justice can have cases involving member states, E.U. institutions, or individuals affected by E.U. regulations.

The infrastructure associated with the European Union is much more comprehensive and complex than for NAFTA. This is indicative of the more encompassing integration that takes place in a customs union compared with a free trade area. It is anticipated that the E.U. will increasingly take political stands so that the power of its member states will be enhanced. Such a complex union requires many institutions to make sure that the various functions of government are met.

The Common Agricultural Policy (CAP) is the cornerstone of agriculture in the E.U. It was one of the reasons why the six original members unified and was the only common policy among them for many years. The CAP has clearly had huge impacts on production levels, trade status, rural development, and consumer prices in the E.U. Chapter 10 explains the main aspects of the CAP and how it has changed over time. The CAP has always been the main budget item for the E.U., but as common activities expand, agriculture's role will diminish relative to other sectors.

SUMMARY

1. Customs unions (CUs) have no trade barriers among members, and the members also have a common external tariff. Often a CU will have free movement of production factors (labor, investment capital, etc.), harmonized standards, and even a common currency (in the case of the European Union).
2. Free trade areas (FTAs) allow the member countries to keep their own trade barriers with nonmembers, but there are rules of origin that keep nonmember goods from being reexported between FTA members.
3. Preferential trade agreements (PTAs) have possible trade diversion impacts. Trade diversion occurs when the PTA encourages member nations to switch imports from efficient nonmember producers to member producers because of tariffs. The extent of trade diversion depends on the final tariffs relative to their pre-PTA levels.
4. The Canada–U.S. Trade Agreement (CUSTA) went into effect on January 1, 1989, with tariffs between Canada and the United States falling to zero over a period of five to ten years. The negotiations were very smooth, but the effects have been significant for some agricultural products, particularly higher-valued agricultural products.
5. The North American Free Trade Agreement (NAFTA) became effective on January 1, 1994, through bilateral agreements among Canada, Mexico, and the United States. The negotiations were much more difficult, and three side agreements had to be signed prior to the final enactment. The environmental side agreement established the Commission on Environmental Cooperation to ensure that environmental concerns are addressed. The agreement has a five- to fifteen-year transition period, and many agricultural items have tariff-rate quotas.
6. The Mexican peso crisis in late 1994 distorted NAFTA effects during the initial two years, but the crisis was resolved by 1997, thanks to U.S. financial guarantees that supported the Mexican banking sector.
7. The move toward a united Europe began in 1957 with the six-member European Economic Community. Over time, the confederation has moved to a customs union with an enlarged membership. The Single Europe Act of 1987 committed the European Union (E.U.) to harmonized standards by 1992. Monetary union, and the launch of the euro, came on January 1, 1999.

QUESTIONS

1. Should the United States move toward expanding the NAFTA to include countries such as Chile, Argentina, and other Latin American nations? Why?
2. Give examples where the NAFTA resulted in trade diversion. How would you estimate the losses from trade diversion?
3. Consider one of the manufacturing plant closings in your area. To what extent was that closing due to NAFTA? To what extent was it due to general globalization?
4. How will the entry of new countries, particularly countries from Central and Eastern Europe, influence the E.U.'s governance and politics?

5. How has agricultural trade between the United States and Mexico changed since the NAFTA? Use up-to-date numbers.

REFERENCES

Ballenger, Nicole, and Barry Krissoff. "Environmental Side Agreements: Will They Take Center Stage?" In Bredahl, Ballenger, Dunmore, and Roe (Eds.), *Agriculture, Trade, and the Environment: Discovering and Measuring the Critical Linkages.* Boulder, CO: Westview Press, 1996.

Bhagwati, Jagdish, and Arvind Panagariya. *The Economics of Preferential Trade Agreements.* Washington, DC: American Enterprise Institute, 1996.

————"Preferential Trading Areas and Multilateralism—Strangers, Friends, or Foes?" In Bhagwati and Panagariya (Eds.), *The Economics of Preferential Trade Agreements,* Washington, DC: American Enterprise Institute, 1996.

Congressional Budget Office (CBO). "A Budgetary and Economic Analysis of the North American Free Trade Agreement." Washington, DC: Government Printing Office, July 1993.

The Economist, October 25, 1995.

International Monetary Fund (IMF). "International Financial Statistics." Various Issues.

Munirathinam, Ravichandran, Michael Reed, and Mary Marchant. "Effects of the Canada–U.S. Trade Agreement on U.S. Agricultural Exports." *International Food and Agribusiness Management Review.* Volume 1, No.3 (1998): 403–15.

Nelson, Frederick, Mark Simone, and Constanza Valdes. "Comparison of Agricultural Support in Canada, Mexico, and the United States." Agricultural Information Bulletin No. 719. Economic Research Service, U.S. Department of Agriculture, September 1995.

Runge, Ford. *Freer Trade, Protected Environment: Balancing Trade Liberalization and Environmental Interests.* New York: Council on Foreign Relations Press, 1994.

Tweeten, Luther, Jerry Sharples, and Linda Evers-Smith. "Impact of CFTA/NAFTA on U.S. and Canadian Agriculture." International Agricultural Trade Research Consortium Working Paper, 1997.

Chapter 8

Macroeconomics and Its Influence on International Trade

An important concept that has been omitted from the analysis to this point is the exchange rate. All the supply and demand curves presented thus far have, at least implicitly, been denominated in U. S. dollars. This is fine for the United States and can accurately portray trade between regions where there is no exchange of currencies. However, in dealing with international trade, there are usually two currencies involved, thus the exchange rate between those currencies is important.

Farmers in Argentina are interested in the peso price of their wheat, not the dollar price of their wheat. Japanese consumers who purchase apples are interested in the yen price of the apple, not the dollar price. The prices among foreign markets are related, through the exchange rate, but one must be clear that supply and demand functions in foreign countries are influenced by the appropriate currency price, not by the U.S. dollar price. The effects of the U.S. dollar are indirect, through the exchange rate.

This chapter covers important elements of the macro economy, mostly the exchange rate, and their influence on agriculture and agricultural trade. The effects of an exchange rate change on exports and domestic prices are covered for small and large countries. The chapter also covers how exchange rate policies and adjustments differ between fixed exchange rate and flexible exchange rate regimes. Critical linkages among the exchange rate, interest rate, and inflation rate are outlined within the flexible exchange rate system. Finally, the relationship between exchange rates and purchasing power is discussed, along with the relevance of these macroeconomic forces on agriculture.

EFFECTS OF AN EXCHANGE RATE CHANGE

Exchange rate determination is covered later in this chapter, but the exchange rate is a relative price that translates the value of one currency into another. It determines the purchasing power (at least for tradeable goods) of one currency versus another. Throughout this chapter and the rest of the book, the exchange rate, k, will be denominated in foreign currency units per dollar. This is the way that most exchange rates are quoted in the financial literature (e.g., 125 Japanese yen per dollar).

When the exchange rate is defined in this way, k will increase as the dollar increases (appreciates) in value because it takes more foreign currency to purchase a dollar. For example, if the Japanese yen changes from 100 yen per dollar to 125 yen per dollar, the dollar has appreciated in value. In contrast, when k decreases, the dollar is falling (depreciating) in value because it takes less foreign currency to purchase a dollar. This relationship is often difficult to understand initially, so one must be careful in talking about changes in currency values and move slowly at first—and keep a consistent definition for the exchange rate (k).

If there is perfect competition, there are no trade barriers, and transportation costs are zero, then the Law of One Price will hold and the exchange rate will perfectly transfer prices from one country to another. For instance, if a U.S. company is exporting, the price in the foreign country, P_f, will be the price in the United States, P_{us}, multiplied by the exchange rate:

$$P_f = P_{us}k \tag{8.1}$$

Note that the relationship is multiplicative, so that a 10 percent decrease in the exchange rate (which means the dollar has depreciated by 10 percent) will decrease the price in the foreign country by 10 percent (assuming that the U.S. price is unchanged, which is the small country assumption for the importer). Even though the dollar price has not changed, the depreciation of the dollar decreases the price of the product in the foreign market.

The fact that foreign supply and demand functions are denominated in another currency does not cause problems in the analysis until the exchange rate changes. One can think of the foreign supply and demand functions as simply multiplied by a fixed number in order to convert them into U.S. dollars. The supply and demand functions will have the same shapes when they are converted into dollars, so they can be depicted as they have been in past chapters. However, when the exchange rate changes, these curves shift because their conversion into dollars changes.

Figure 8.1 shows how a foreign supply curve is transformed into dollars so that they can be compared across countries. The left panel of Figure 8.1 shows a supply curve in French francs, the relevant currency for French farmers. It has a positive slope, and notice the quantity units associated with the supply curve. Assume that the exchange rate is 6 French francs per dollar. When the curve is transformed into U.S. dollars, only the y axis changes; the x axis (quantity axis) remains the same. In effect, the supply curve shifts down to reflect the fact that the French franc is worth less than the U.S. dollar.

Figure 8.2 shows the impacts when the French franc changes from 6 francs per dollar to 5 francs per dollar. Note the nonparallel upward shift in the supply curve denominated in dollars (to S') if the franc appreciates to 5 francs per dollar. The French are willing to supply only 20 units if the price is $4.80 (24/5) and only

FIGURE 8.1

Transforming a French supply curve denominated in francs to one denominated in dollars.

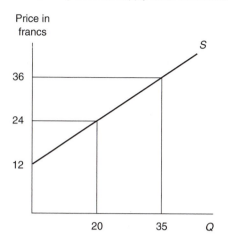

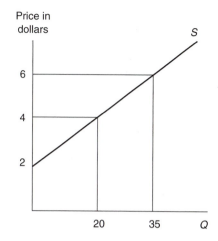

FIGURE 8.2

Effects of an exchange rate change on the French supply curve denominated in dollars.

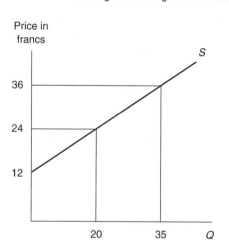

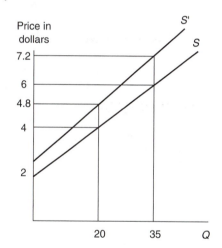

35 units if the price is $7.20 (36/5). The dollar price translates into a lower franc price if the exchange rate is 5 francs to the dollar, so the French supply less. If the franc lowered in value, say to a price of 7 francs per dollar, the supply curve would shift downward in Figure 8.2.

One can perform a similar analysis for demand curves and exchange rate changes. When the franc changes in value from 6:1 to 5:1, the French demand curve denominated in dollars will also shift upward. The appreciation in the value of the French franc reduces the good's price in France, and the people there naturally consume more. This graph is left to the interested reader.

FIGURE 8.3
The effects of a franc appreciation on the excess supply of France.

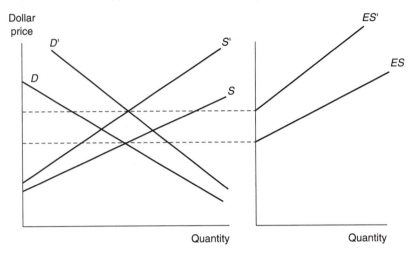

The effects of an appreciation in the value of the French franc are indicated in Figure 8.3, where the French supply and demand curves are shown in dollars. When the French franc changes from 6 francs per dollar to 5 francs per dollar, the supply and demand curves (when denominated in dollars) shift upward to S' and D'. The supply and demand curves are unchanged when denominated in French francs, but they shift when they are transformed into dollars. The upward shift in the supply and demand curves causes an upward shift in the excess supply curve (to ES'). If France is a small country, the dollar price of the good will not change (the ED' curve France faces is perfectly elastic), but French exports will decrease from Q_0 to Q_1, as shown in Figure 8.4.

Thus, when the franc appreciates, the French supply and demand curves will shift upward (relative to where they were with the previous exchange rate). The French will supply less because their price (in francs) has fallen, whereas French buyers will purchase more. The shifts will not be parallel shifts because the relationship in Equation (8.1) is multiplicative: a 10 percent decrease in k (an appreciation in the value of the French franc) will decrease P_f by 10 percent.

Figure 8.5 traces the effects of the exchange rate change back to the dollar-denominated supply and demand curves for France. When the exchange rate changes, the quantity supplied will fall from Q_1 to Q_2 and the quantity demanded will increase from C_1 to C_2. Note that exports decrease from $Q_1 - C_1$ to $Q_2 - C_2$. The U.S. dollar price has not changed, but the franc price has fallen, causing these changes.

If the small country assumption is relaxed, the French face a downward-sloping excess demand for their exports. The appreciation in the French franc causes a sizable decrease in world supply, so the dollar price of the good must increase from P_0 to P_1, as shown in Figure 8.6. The decrease in world trade is less than under the small country assumption because there is a secondary impact of being a large country. The franc price does not go down by the full amount of the exchange rate

FIGURE 8.4
Effects of a franc appreciation on
French exports—small country.

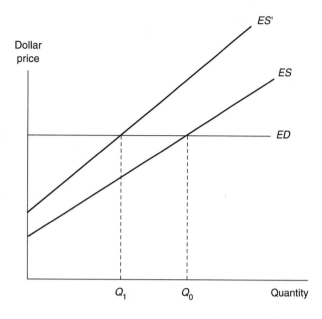

FIGURE 8.5
The effects of a franc appreciation on production and consumption.

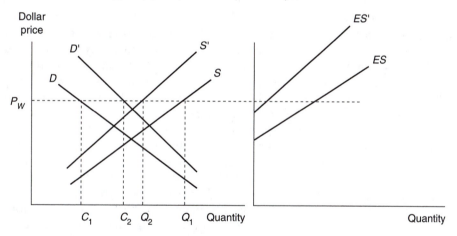

change because the world price is higher. It is clear, though, that world price increases and French exports decrease, so the ultimate franc price is lower than before the devaluation.

A similar analysis can be performed with a depreciation in the French franc or an appreciation or depreciation in the value of the dollar. Note, however, that a 10 percent appreciation in the franc is identical with a 10 percent depreciation in

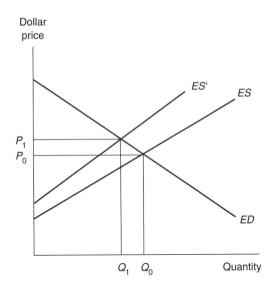

FIGURE 8.6
Effects of a franc appreciation on French exports—large country.

the dollar; and a 10 percent depreciation in the franc is identical with a 10 percent appreciation in the dollar. In general, it is easier to visualize with a three-panel diagram for the country in question and the rest of the world. As an exchange rate changes, one of the panels will have its supply and demand curve shift in a non-parallel fashion to reflect the exchange rate change. Effects on that country and the rest of the world can be traced through the three-panel diagrams.

Exchange rate changes clearly affect the price of tradeable goods. One can see from above that an appreciation in a country's exchange rate decreases the price of its exportable goods and increases the price of importables for its trading partners. The appreciation in the country's exchange rate will also decrease the price of its importables and increase the price of the exportables from its trading partners (if the country whose currency depreciated is "large"). When these effects are added together, an appreciating currency will lead to lower inflation because tradeable goods decrease in price, while a depreciating currency will lead to higher inflation.

There are winners and losers to exchange rate changes for a country. When a country's currency appreciates, consumers are winners because imported goods are cheaper for them. If they travel internationally, their domestic currency will buy more foreign currency, so prices internationally will be cheaper for them. Any domestic firms that export or compete with imported goods clearly lose from an appreciated currency because the domestic prices those firms face are lower.

When a country's currency depreciates, consumers lose because importables and exportables are more expensive. It is also more expensive for people to travel to foreign destinations. Businesses that import goods are losers because their imports also cost more. Yet businesses that are involved in exports or produce goods that compete with importables are clear winners. They face higher prices for their output (in the case of exporters) or face competitors that have higher costs (in the case of import-competing businesses), which is beneficial to their businesses.

FIGURE 8.7
Supply and demand for dollars.

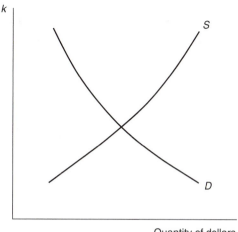

Quantity of dollars

In today's world, when such a large percentage of our economy depends directly or indirectly on trade, the influence of the exchange rate has heightened greatly. Yet the ultimate determination of exchange rates is still difficult for economists to pinpoint. This has become increasingly clear over the past few years, with currency crises happening with increased regularity. The next section deals with factors that help determine exchange rates. The section begins with a historical perspective, when exchange rates were fixed, and moves to the present system, where many exchange rates are allowed to fluctuate freely.

DETERMINANTS OF EXCHANGE RATES

An exchange rate is simply another price that must be determined within the economic system. There is a supply and demand for a currency, and the exchange rate must be related to those curves. Figure 8.7 shows a supply and demand curve for U.S. dollars. One can think of these as supply and demand curves facing the French central bank as people exchange their francs into dollars for international transactions. Let us assume that there are only two countries in the world, France and the United States, so if the French want to make an international transaction, the transaction must involve dollars. However, the French people always use their central bank to buy and sell their dollars.

The demand for U.S. dollars comes from French people who wish to purchase U.S. goods and services or invest in U.S. assets (bonds, stocks, or fixed plant and equipment). These people go to the French central bank to obtain these dollars that they need by exchanging their francs. The supply of U.S. dollars comes from French people who have sold their goods, services, or assets to Americans. These French people take the dollars they have earned to the French central bank and exchange them for French francs.

The international exchange system can work in two basic ways: under fixed exchange rates or under flexible (floating) exchange rates. Each system has positive

FIGURE 8.8
Effects of exchange rate on central bank reserves.

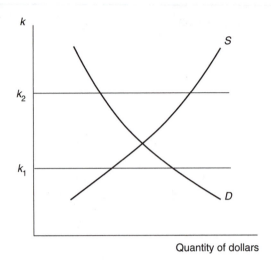

and negative aspects, but today's flexible system is the only one that can accommodate the huge daily volume of currency transactions (at least for the major developed countries). However, to better understand the concepts of exchange rate determination, it is best to begin with the fixed exchange rate system.

The Bretton Woods exchange rate system, named after a town in New Hampshire where the system agreement was signed, was the basic fixed exchange rate system that began after World War II. Here the U.S. dollar was fixed relative to the price of gold, and other countries agreed to defend (peg) their currencies relative to the U.S. dollar. This is called a *gold exchange standard*, and it results in fixed exchange rates for all currencies. Each country would hold international reserves (usually in the form of gold, U.S. dollars, or other international currencies) that it would use to buy or sell its currency to keep its value fixed. The central bank in each country was charged with the responsibility of keeping the exchange rate at the announced level.

Assume that k_1 is the fixed exchange rate between the French franc and the U.S. dollar. If the supply and demand situation was as portrayed in Figure 8.8, the demand for dollars would exceed the supply. The French central bank would need to supply U.S. dollars by purchasing the francs that were offered by people. The fixed exchange rate is defended by the willingness of the central bank to supply or purchase U.S. dollars at that exchange rate. The situation in Figure 8.8 depicts a balance of payments deficit because the French central bank is supplying its dollar reserves to the market. At this point, one could think of the French franc as overvalued at that moment because the demand for dollars exceeded their supply (with the difference made up by the central bank buying francs).

There can be problems if the currency is overvalued for a long period of time because dollar reserves are limited, making a persistent overvaluation impossible unless the country borrows reserves from others (international agencies, foreign governments, etc.). When there are persistent outflows of international reserves in a fixed exchange rate system, the country is usually forced to devalue its currency (change the exchange rate to reflect more accurately the supply and demand situation).

If the fixed exchange rate was k_2 instead of k_1, the French central bank would be accumulating U.S. dollars because people have received more dollars from exporting or selling their assets to Americans than is required to import goods or invest abroad. The French people want to exchange their U.S. dollars for French francs, and they do it at the French central bank. In this situation, France is experiencing a balance of payments surplus because the central bank is accumulating dollars. The franc is, at least temporarily, undervalued, and the central bank will see its dollar reserves increasing. This temporary undervaluation of the franc does not cause problems as long as the French central bank doesn't mind accumulating international reserves.

Currency depreciations in a fixed exchange rate system normally bring about a fair degree of pain because the depreciation is usually large (10 percent or more) and is often not well anticipated. Because exporters, importers, and investors have developed business plans that incorporate the fixed exchange rate, a large change in its value can be detrimental to profitability. So, a fixed exchange rate system is fine as long as countries have enough foreign reserves to defend their currency. However, when they run low on foreign reserves, the trauma associated with the currency depreciation can be great.

The Bretton Woods system lasted from 1945 to 1973. In the early 1970s, the stability of the system was compromised by U.S. inflation and the decline in the U.S. dollar as an international currency. The United States attempted to stabilize the system through devaluations in 1971 and 1973, but decided to let its currency float in late 1973 (it no longer fixed the U.S. dollar's value relative to gold). Without the U.S. dollar fixed to gold, other countries saw no purpose to fix their currencies relative to a dollar that fluctuated on a daily basis. Since 1973, the world has been off a fixed exchange rate system.

FLOATING EXCHANGE RATE SYSTEM

There are no balance of payments surpluses or deficits with a purely floating exchange rate because quantity demanded and quantity supplied for currencies are equated by the exchange rate. The central bank does not need to buy or sell currencies for the market to clear. However, the currency's value will fluctuate so that exporters, importers, and investors will not know what a currency's value will be at any point (with certainty). The risk associated with international transactions has increased because exchange rates fluctuate, but there is less need for the country to hold large international reserves or to have large currency devaluations.

If the only need for a currency is for trading goods, understanding exchange rate fluctuations with a floating rate is relatively easy. The demand for French francs would come from international customers wishing to purchase French exports, whereas the supply of French francs would come from international suppliers selling products in France (French imports). The supply and demand functions for French francs would be straightforward, and adjustments in

Basic Exchange Rate Systems

There are three basic exchange systems for countries in the world today: the free floating, managed floating, and pegged systems. The free floating system—which is followed by the United States, Canada, and Japan—allows exchange rates to fluctuate freely on a daily basis. There may be some government (central bank) buying and selling of currencies, but these transactions are only occasional. The managed floating system, which was used by European Union countries leading up to the euro, is where the central bank has agreed to defend upper and lower bounds on the exchange rate. If the exchange rate hits the upper or lower bound, the central bank will engage in currency transactions to defend the currency. Pegged exchange rate systems, which are used by more countries than the other two systems, are when a country's central bank defends the currency for a fixed rate relative to some other currency (usually the U.S. dollar or the French franc). Pegged systems have fixed and flexible elements to them because the country's currency is fixed relative to one currency but fluctuates relative to all other currencies.

the franc would be easy to explain. If the demand for French exports exceeds the supply of French imports, then the franc's value will increase. This increase in the value of the franc will make exports more expensive for the French and imports less expensive, but these effects are easy to document and depend on rather predictable factors such as supply and demand elasticities for goods.

The demand for currencies has never been exclusively for goods in the modern era, but until recently, exchange rates have been mostly influenced by trade in goods. This changed drastically in the 1970s, and since that time world currency trade has changed from a goods phenomenon to a monetary phenomenon. Most currency is traded for investment purposes, not trade purposes. The huge expansion in currency futures and the increased flow of money among countries to purchase bonds, stocks, and other real assets have drastically changed the nature of currency markets. International trade does not make currency values move, but interest rates and investment opportunities between countries do. It is estimated that the trade in currencies for investment purposes is over twenty times the trade in currencies for trade purposes.

In an environment where currency values are determined by investment flows, the interplay among interest rates, exchange rates, and inflation rates is critical. Interest rates and inflation rates are linked through the Fisher equation, where the nominal interest rate, i, is equal to the real interest rate, r, plus the expected rate of inflation, Δp:

$$i = r + \Delta p \tag{8.2}$$

Balance of Trade Deficits—Politics or Economics?

One might wonder why there is such a fixation on the balance of trade if most currencies are demanded for investment purposes. The news media still report the merchandise trade balance each month and insinuate that a balance of trade deficit is bad. In the current economic environment (with floating exchange rates and no balance of payments deficits), a balance of trade deficit does indicate that the country is importing more goods than it is exporting, but this should not be worrisome by itself. A balance of trade deficit is more reflective that foreigners are demanding the country's currency for investment purposes and that there is a net inflow of money into the country for investment (foreign investors are purchasing stocks, real estate, bonds, and other assets). This inflow might be because savings rates are low or because investment opportunities are great. If the capital inflows are due to a low savings rate, the country might want to take measures to increase savings. If the capital inflows are due to excellent investment opportunities, the country should be pleased with the trade deficit.

The real rate of interest reflects the return that an investor receives for postponing consumption over some time period. The real rate is obtained by subtracting the expected rate of inflation from the nominal rate because the purchasing power of money has fallen. The real rate of interest tends to be common across countries, but inflation rates differ markedly. Thus, *countries with higher nominal interest rates should also have higher inflation rates.*

This is borne out by a simple examination of interest rates and inflation rates among countries. Table 8.1 (taken from *The Economist*) shows the relationship between interest rates and inflation rates for selected countries. Notice that the countries with a higher past inflation rate have a higher interest rate. Remember, however, that the relationship is forward-looking: present interest rates reflect expectations of future inflation. China is a case where investors do not think the most recent inflation performance (−1.2 percent) is indicative of longer-term inflation. Japan is a special case where nominal interest rates have dropped as far as possible and future deflation is a real possibility.

LINKAGES BETWEEN INTEREST RATES AND EXCHANGE RATES

Nominal interest rates among countries are linked because investors can purchase government bonds or other debt instruments for any leading developed country with relative ease. However, the exchange rate must enter the investment decision because anyone who purchases a bond outside their country must

Developed Countries

	i	Δ*p*
Euro - 11	3.1%	0.8%
Japan	0.1	−0.2
U.S.	4.7	1.7

Less Developed Countries

	i	Δ*p*
Chile	8.2%	3.8%
China	12.0	−1.2
Poland	13.1	6.8
S. Korea	6.6	0.2
Turkey	80.0	63.9
Venezuela	34.5	29.5

Source: *The Economist*, March 13, 1999.

exchange currencies. If an American investor wants to purchase a French government bond, that investor must change dollars into francs. After the bond matures (or is sold), the American would presumably want to change the francs back into U.S. dollars. Thus, the future change in the exchange rate between the U.S. dollar and French franc will impact the American investor's true return on the French bond.

No one knows future exchange rates with certainty. However, if there is no expected change in the value of the franc relative to the dollar, the interest rate on U.S. government bonds should be identical with the interest rate on French government bonds (as long as there are identical default risks). If U.S. interest rates were higher than French interest rates (and there was no anticipated exchange rate change), then everyone would choose to invest in the United States to receive the higher interest rate. Any anticipated depreciation in the value of the French franc should be matched by a higher interest rate on French bonds to compensate for the capital loss in the bonds' value (in dollar terms). An example should help explain this situation.

Suppose an American investor had $100,000 that she wished to invest in government bonds for a one-year period. Assume that the investor is deciding whether to purchase American or French government bonds and that the current exchange rate is 6 francs per dollar. Assume further that the French bonds carry an interest rate of 6 percent and American bonds carry a rate of 5 percent, both bonds mature in one year, and each bond has no default risk. The situation is depicted in Table 8.2, where the left column shows the return calculations for the American bond and the right column shows the return calculations for the French bond.

TABLE 8.2   Comparing Bond Returns between Countries

	American Bond	French Bond
Face amount	$100,000	600,000 francs
Interest return	5,000 (5%)	36,000 (6%)
Bond's value after one year	105,000	636,000

The American calculations are very straightforward. The $100,000 earns an interest rate of 5 percent, so the bond is worth $105,000 upon maturity. The French calculations are different because the French bond is denominated in francs, so the American investor purchases a French bond worth 600,000 francs. After one year this bond is worth 636,000 francs, but the key question is, What is the bond's worth in dollars upon maturity? Without knowing the future exchange rate, there is no way to know the dollar return on the French bond with certainty.

If there is no change in the exchange rate over the year, the French bond would be worth $106,000, resulting in a 6 percent return in U.S. dollars after one year. However, if the French franc fell in value, the return would be less than 6 percent. For instance, if the franc fell to 6.36 francs per dollar, the bond would have a value of only $100,000 at the year's end and the return would be 0 percent. If the French franc appreciated in value, the return would be above 6 percent. For instance, if the franc increased in value to 5.5 francs per dollar, then the bond would be worth $115,636 after one year, generating a return of over 15.6 percent. Informed investors who wish to maximize their return (in dollar terms) must incorporate expected exchange rate changes into their portfolio decisions.

Obviously, the decision on which bond to purchase depends crucially on what exchange rate is expected to prevail in the future. Fortunately for investors, futures markets exist for major international currencies, which allow investors to hedge their transactions by "locking in" a forward exchange rate. One can purchase or sell any major currency for up to one year in the future, so the future exchange rate can be determined for investment purposes. The forward exchange rate allows investors to lock in the future exchange rate with certainty and reduce the risk associated with investment strategies.

Since American government bonds and French government bonds have identical risks, their expected returns should be identical. If investors could lock in a higher return on French government bonds, then no one would purchase American government bonds and American interest rates would be forced to rise. Likewise, if investors could lock in a higher return on American government bonds, then no one would purchase French government bonds. In equilibrium, their expected returns must be equal. This logically means that *any difference in the interest rate on different bonds (of equal risk) between countries must reflect expected changes in the exchange rate*:

$$i_{us} = i_f - e(\Delta k) \tag{8.3}$$

Equation (8.3) is called the international Fisher equation. The interest rate in the United States, i_{us}, is equal to the foreign interest rate, i_f, minus the expected depreciation in the foreign exchange rate (if the currency depreciates, the dollar return will be lower), $e(\Delta k)$. Relative to the U.S. interest rate, foreign investors make money off either the interest rate or the increase in value of the foreign currency. Therefore, a simple comparison of interest rates among countries will indicate which currencies are expected to appreciate and which are expected to depreciate in the future.

If there is a linkage between nominal interest rates and inflation for each country (Equation 8.2), and there is a linkage among countries between nominal interest rates and expected exchange rate changes (Equation 8.3), there must be a linkage between expected exchange rate changes and inflation. *Countries with high inflation rates will have currencies that are expected to depreciate (become less valuable) in the future.* This is one version of the purchasing power parity (PPP) condition: the rate of exchange rate change is equal to the difference in inflation rates between two countries. This makes sense because countries with higher inflation rates have currencies that are becoming less valuable relative to goods, so these currencies should also become less valuable relative to currencies that are more stable. Another view of Table 8.1 is that the countries with high nominal interest rates and inflation rates are the same countries where the investment community expects a depreciation in the currency's value.

The relationships between interest rates, inflation rates, and exchange rates are key considerations for central bank monetary policies throughout the world. It is well known that expansionary monetary policy in today's globally integrated financial world will ultimately increase a country's inflation rate, increase nominal interest rates, and result in currency depreciations. Thus, if a central bank wishes to keep its currency stable with low interest rates, it must fight to make sure that inflation is in check.

PURCHASING POWER PARITY AND OVERVALUED EXCHANGE RATES

Exchange rate changes, especially on a short-term basis, are brought about by monetary phenomena, but there is still much interest in the exchange rate and its impact on international trade. Many companies rely heavily on international markets for their products, and farmers in particular have products that rely on international markets for a large percentage of their output. The exchange rate might not be primarily determined by agricultural trade, but the exchange rate certainly affects agricultural trade volume and agricultural prices.

Because exchange rates are determined by monetary policy and the relationship between monetary policy and interest rates is not well known, subject to errors and time lags, one sees exchange rates fluctuating a great deal over time. This can be disturbing because one does not know exactly what exchange rates will be in the long term. Without such knowledge, it is difficult to make wise investment decisions that involve international trade or investment. For this reason, economists have tried to come up with longer-term theories of exchange rate determination so that one can make better decisions and compare current exchange rates with these "long run" exchange rates to predict longer-term currency movements. If an exchange rate is below its "long run" value, it is undervalued, and if it is above its "long run" value, it is overvalued.

The most common view of long-term currency movements is the purchasing power parity (PPP) theorem, which was referenced earlier. This theorem has many versions, but ideas from some versions are commonly used in the business literature today. The absolute version of the PPP theory states that in the long run, currencies will reflect their relative purchasing power among countries. If absolute PPP holds, then the cost of a basket of goods would be the same in all countries (when the prices are translated into a common currency).

Absolute PPP is related to the Law of One Price and is not generally accepted by economists. Even in a world of free trade and no transport costs, absolute PPP would not hold because not all goods are tradeable. Housing and many services, for example, are not generally traded between countries, so there is no pressure for these prices to equalize among countries. Despite these shortcomings, ideas associated with absolute PPP are still commonly used.

Periodically, *The Economist* will publish its "Big Mac" index to estimate whether currencies are overvalued or undervalued relative to the U.S. dollar. The magazine collects information on the price of a Big Mac sandwich from McDonalds in countries throughout the world and converts the price into U.S. dollars using the current exchange rate. If the dollar price is above the U.S. price, the currency is said to be overvalued, whereas if the dollar price is below the U.S. price, the currency is said to be undervalued. *The Economist* makes it clear that there are problems with this index, but argues that the Big Mac is a standardized product sold throughout the world, so it provides a legitimate estimate of relative currency valuations.

Despite the existence of trade barriers, nontradeable goods, and other price rigidities among country markets, one would still expect prices to move together among countries and that prices for tradeable goods should be nearly equal if exchange rates are at free market levels. Instead of absolute PPP, though, economists are more comfortable with the relative PPP theorem, which states that currency changes between countries should reflect relative inflation rates. This is identical with the proposition espoused earlier, that countries with high inflation rates and nominal interest rates will have depreciating currencies. These relationships tend to be preserved because of portfolio investment flows among countries.

MACROECONOMIC FACTORS AND U.S. AGRICULTURE

The value of the dollar relative to other currencies is very important for U.S. agriculture because the United States exports such a high percentage of its output. The prices of agricultural commodities and the volume of U.S. exports increase when the value of the dollar falls, which is certainly good for U.S. agriculture. International customers can purchase more U.S. agricultural products with the same amount of their currency as the dollar depreciates.

DISCUSSION BOX 8.3

The Asian Financial Crisis and U.S. Agriculture

The financial crises in Indonesia, Korea, Malaysia, and Thailand highlight the effects that exchange rates can have on U.S. agricultural exports. Between July 1997 and July 1998 the value of the Indonesian rupiah, Korean won, Malaysian ringgit, and Thai baht fell by 70 percent, 35 percent, 34 percent, and 34 percent, respectively. At one point during that year the rupiah had fallen by 85 percent (going from 2,400 rupiah per dollar in July 1997 to 16,000 rupiah per dollar in January 1998). The reasons for the financial crisis are complex and beyond the point of this discussion. However, the crisis of economic and political confidence in these countries caused severe hardships on their banks, companies, and citizens.

These financial waves were felt not only on Wall Street, through fluctuations in stock prices, but throughout U.S. agricultural markets, because of reduced exports to these destinations. In 1996, Korea imported $3.8 billion in U.S. agricultural products, but it imported only $2.2 billion in 1998, a 42 percent drop. The drop in Malaysian imports of U.S. agricultural products from 1996 to 1998 was 54 percent; it was 47 percent in Indonesia and 29 percent in Thailand. In total, the value of U.S. agricultural exports to these destinations declined by $2.5 billion in those two years. The Asian financial crisis, which also affected exchange rates for countries throughout Asia and Latin America, was a major reason that U.S. agricultural exports peaked in 1996.

The flip side of the coin is that U.S. agriculture suffers when the value of the dollar increases relative to other currencies. If the United States is a residual supplier, as a result of government storage programs or other policies, U.S. exports can fall dramatically as the dollar's value rises. This was in evidence during the early and mid-1980s and has also been important in recent years when many countries have experienced currency crises and steep declines in their currency values (meaning a steep increase in the dollar's value).

There is not much that U.S. agriculture can do to reduce the impacts of exchange rate changes, since exchange rate movements are virtually independent of the trade in goods. One can only hope that as the international community learns more about financial crises, good banking practices, and macroeconomic policies, the fluctuations in the exchange rate will be reduced over time.

SUMMARY

1. The exchange rate is the price of one currency versus another. Exporting firms like to have a lower valued currency because it makes their products less expensive for foreign buyers.

2. When the country is large, a change in the exchange rate causes a nonparallel shift in the excess demand or excess supply curve. In this case, the world price will also change as a result of the exchange rate change.

3. Under a fixed exchange rate regime, countries will run balance of payments surpluses or deficits, so they must hold foreign currency reserves to cover the deficits. Under a flexible exchange rate system, there are no balance of payments deficits, but the exchange rate will change on a daily basis.

4. In the current environment, capital flows determine exchange rate changes, so the linkages among the exchange rate, interest rate, and inflation rate are critical. International capital movements will ensure that the expected net rate of return in holding different currencies will be identical among countries of similar risks.

5. Countries with higher nominal interest rates should also have higher inflation rates. Any difference in the interest rate on different bonds (of equal risk) between countries must reflect expected changes in the exchange rate. Countries with high inflation rates will have currencies that are expected to depreciate (become less valuable) in the future. These movements in exchange rates are consistent with the purchasing power parity theory.

QUESTIONS

1. Argentina's short-term interest rate is 7.2 percent, while Venezuela's is 34.5 percent . What can you say about macroeconomic conditions in the two countries?

2. Why is it impossible for most developed countries to follow a fixed exchange rate system in today's world economy?

3. Why would you expect the purchasing power parity to hold between two countries?

4. If you are a horse farm owner in central Kentucky, would you want a strong or weak U.S. dollar? Why?

5. I was with an engineering professor at a bank in Kuala Lumpur in April 1997. The interest rate on savings deposits posted at the bank was 14.5 percent. My friend suggested that we should deposit all our money in the bank. What other information did he need to know? How would our rate of return compare with U.S. returns?

Chapter 9

Trade and the Environment

The increased integration of the world economy through freer trade and capital flows has led to friction among countries because of differing domestic regulations. Even when border controls on capital and trade are low, differing domestic regulations can have impacts that go beyond the border of a country. Environmental regulations are a classic case of this conflict, and these regulations have been a major source of discussions and conflicts since the early 1990s. Increased emphasis on domestic regulations and their impact on trade comes at a time when the attention afforded the world's environment is also increasing. The intersection of these two trends, globalization and concerns about the environment, has focused attention on their interaction.

Environmental regulations are not the only domestic policies that have trade consequences. Earlier chapters examined agricultural price and income policies, which were addressed through the Uruguay Round of the GATT. Examples of other domestic policies that have trade consequences include antitrust laws and labor standards, which impact the cost structures of domestic companies and, therefore, can have impacts on trade patterns. Countries have vested interests not only in their own domestic policies but also in the policies of their trading partners. Increased economic integration among countries has led to conflicts on these domestic issues and a demand for more harmonization, especially concerning environmental policies.

This chapter investigates how issues associated with the environment are influencing trade discussions, negotiations, and policies. Trade and the environment is a crucial issue these days because incomes are higher and there is an increased demand for environmental standards as income increases.[1] There are also many examples where parts of the earth have passed their ability to assimilate and process pollution. The horrible effects of what environmental degradation can do to an economy are seen in chemical runoff in Eastern Europe and the former Soviet Union, smog in Los Angeles, the hole in the ozone layer, and the like. These and

---

[1]The World Development Report, 1992, is the major source quoted for this conclusion.

The Relationship between Income and Pollution

Grossman and Krueger found that there is an inverted U-shaped relationship between income and pollution that peaks at an income level of $5,000 per capita. Thus, for countries with a per capita income above $5,000, increased incomes will increase the demand for environmental services. This is a common finding by researchers, though there is much debate on the exact income level associated with the pollution peak.

other examples have shown that it is possible to severely damage the environment, cause tremendous human suffering, and increase the long-term risk of diseases.

The discussion here focuses on the conflict between the free trade community and those concerned with the world's environment, how this conflict interacts with the GATT accord, what economic analysis can say about trade and environmental issues, how these conflicts were handled during the NAFTA and Uruguay Round negotiations, and finally how the conflict between trade and the environment can be resolved in the future.

Environmental externalities exist because private consumption or production has external effects on others (e.g., when a plant discharges polluted water as part of its production process). In this case, private costs or benefits are different from social costs or benefits, and there are spillover effects on society. One can categorize three forms of environmental externalities: local, transnational, and global. Local externalities affect only the country where the production or consumption takes place, and one typically allows the country in question to resolve these problems. Transnational externalities happen when production or consumption in one country has spillover effects into a bordering country. Polluted water might ultimately enter a foreign country, and so some of the social costs associated with the pollution are borne by another country. Global externalities happen when production or consumption in one country has spillover effects on all other countries. An example would be using chlorofluorocarbons (CFCs) that are known to destroy the ozone layer of the earth's atmosphere.

There are a number of multilateral agreements dealing with environmental issues that have trade consequences. The first one was the Convention Relative to the Preservation of Fauna and Flora in their Natural State (1933), which regulated trade in natural species and trophy animals. Other well-known agreements include the International Plant Protection Agreement (1951), the Convention on International Trade in Endangered Species (1973), and the Montreal Protocol on Substances That Deplete the Ozone Layer (1987). Each of these agreements addresses global environmental issues associated with production or consumption by establishing standards and policies that signatories pledge to follow. These agreements were worked out focusing on the environmental aspects of the agreement, then making sure that commercial arrangements support the agreement and protocols.

DEFINING THE CONFLICT BETWEEN TRADE AND THE ENVIRONMENT

There are conflicts between trade and the environment for three reasons. First, trade and trade policies impact the environment because they change production and consumption for countries. Trade (and trade liberalization) allows countries to specialize in production of those products that fit its resource base. Trade liberalization will change not only income and consumption levels but also production location. If the changes in consumption or production generate environmental externalities, trade itself might be viewed as leading to environmental degradation.

Some of the environmental problems associated with increased trade come about because property rights are not well established for some natural resources. If property rights are not well established, then producer and consumer behavior is not efficient from a social point of view. Disputed, ambiguous, or nonexistent property rights are a major problem throughout the world. If residents of the rain forest in Brazil view the land as theirs, then they aren't concerned that others are upset when the forest is cleared so that the residents can plant crops. However, if the world views the rain forest as a vital global resource that everyone owns, then they expect rain forest residents to behave differently.

Often property rights are well known, but very difficult to enforce. The Montreal Protocol agrees that CFCs are a detriment to the world's ozone layer. Ozone depletion will result in increased cancer incidence throughout the world. Despite the concurrence on the necessity of action, a key question is, How can this agreement be enforced? An answer to many of these environmental debates is to make producer and consumer costs and benefits coincide with social costs and benefits.

A second conflict between trade and the environment comes about because environmental policies in one country can impact other countries through trade. Each country has its own unique geographic features, assimilation capacities, cultural traditions, income, and other factors. These variations make it reasonable for environmental standards to differ by country. Yet if environmental standards vary by country, this will influence the trade pattern among countries by giving a competitive advantage to firms in one country over another. Furthermore, environmentalists and import-competing businesses say that these environmental policies of individual countries can effectively be overridden by the lower standards of exporting countries if unbridled market access is allowed. They use these arguments to call for harmonized standards. Yet harmonized standards may not be economically efficient because the costs of pollution differ by location.

Often, the debate about environmental standards and regulations turns to a North (rich countries) versus South (poor countries) controversy. The North wants the South to impose tighter regulations on its firms and citizens to preserve the environment. Yet the South sees that most current environmental problems are the result of unchecked economic growth in the North. Why should the South agree to the North's environmental standards when such standards were much more lax for the North many years ago (when the North's economic status was more comparable to the South's economic status today)? The South sees this attitude, where the North wants to dictate environmental standards to the South as eco-imperialism. The South also sees high environmental standards as a way to keep them poor by restricting their access to markets in the North. Low points out that poverty is the "most aggravating and destructive" environmental problem of all because people do not care about the environment if they are simply trying to survive.

Finally, many people feel that trade policies are a way to force countries into discussions about environmental issues and agree to environmental aspects of trade agreements. There are environmental problems throughout the world, and there is no good international organization that can encourage countries to help solve problems with externalities. Environmental agreements, as referenced earlier, are available for some issues, but they are difficult to negotiate and often their terms are "watered down" to get agreement by large numbers of countries. Trade barriers can be a way to force countries into environmental discussions because these barriers directly affect economic activity. People who argue that trade barriers can be used as a weapon feel that the current multilateral trade organizations, specifically the GATT or WTO, do not deal effectively with issues associated with the environment.

For the purposes of discussion, this chapter will divide the trade and environment conflict between two opposing camps: the free traders and the environmentalists. This is a convenient abstraction from reality because it polarizes the viewpoints but does allow discussion and analysis of the basic tenets of each side.[2] As is normally the case in such polar arguments, the truth (and solution) lies somewhere between the two poles. In fact, the terms *free trader* and *environmentalist* will be used loosely in the chapter to describe a tendency toward these polar views. This is not to confuse the reader but simply to clarify the debate that continues to rage.

An environmentalist is one who believes that economic growth is bad because higher living standards will consume scarce natural resources and generate pollution that will reduce welfare. They suggest that, at a minimum, any increase in incomes through growth is more than overcome by pollution increases. They assert that environmental spillovers are so rampant that trade liberalization should be fought because the changing pattern of production will only serve to increase environmental degradation. Further, they believe trade liberalization will provide an incentive for governments to reduce environmental standards so that they can compete (on a free-trade basis) with countries that have lower environmental standards.

Environmentalists argue that there are market and political factors that keep environmental standards low in many countries. Freer trade will put increased pressures on governments to lower environmental standards to increase their

---

[2]One will find individuals who adhere to the polar views, but in general most people would classify themselves as being between the definition of a free trader and an environmentalist. Esty argues that there are two types of environmentalists: ones who think that economic growth is bad and others who think that trade might not naturally lead to an improved environment. The latter group wants to make sure that some of the gains from trade are used to ensure that growth is sustainable and to preserve public health and ecological resources.

firms' competitiveness and save jobs. This will result in an endless circle of lowered environmental standards and economic well-being. As mentioned earlier, they believe trade policies are one of the few tools available to influence groups outside their country into agreements to improve the environment.

A free trader is one who views the market as the primary solution to economic questions, and the role of government is to facilitate the workings of private markets. Free traders may recognize that environmental problems exist because of externalities, but these problems are very minor and can be overlooked relative to the other economic problems of trade liberalization (or can be addressed through other means that will be discussed later). They assert that environmentalists are really disguised "protectionists" who only want to use the environment as an excuse to protect domestic industries in a global market. They believe the world trading system is already fragile because of persistent protectionism, so allowing trade barriers that compensate for differing environmental standards (or other proposals of environmentalists) will only cloud the debates—where do environmental standards end and comparative advantage begin? Free traders assert that nations should have control over all domestic policies and regulations because nations are sovereign.

There is a clash not only in ideologies between free traders and environmentalists but also in style and cultures (Esty). Free traders are accustomed to closed-door negotiations, which are very secret. Each country has winners and losers for any negotiating stance, and so the offers in any negotiations would be controversial if the positions were made public. Without closed-door negotiations, special interests would gain and consumers would lose. Further, compromise is an important ingredient in all multilateral trade negotiations because of the political realities associated with the changing pattern of production from trade liberalization.

Environmentalists prefer open debates because they believe the closed-door meetings help the large corporations, which can profit from environmental degradation. These corporations would dominate closed-door discussions and be the ultimate winners because the environment would be sacrificed. Society, which is made up of a large number of consumers, must be brought to action through demonstrations, publicity, and meetings because without mobilization of the silent majority, the people would lose out and their voice would be drowned. Compromise is not valued because there is an overarching social good being threatened.

Free traders trust the market to generate wealth because government regulations will only get in the way of creative entrepreneurs. A tax or subsidy scheme to compensate for the difference between private and social costs will encourage economically optimal reactions by consumers and producers. However, the environmentalists prefer laws that establish fixed limits on pollution or establish fixed environmental standards because putting a price on pollution (through taxes or subsidies) is unethical and distasteful. They feel that taxes allow big corporations to buy their way out of their environmental degradation and that paying them subsidies to use pollution-reducing technologies is unthinkable.

There are similarities, however, between free traders and environmentalists. Both argue for causes that face powerful lobbies (labor is typically against free traders, and businesses are typically against environmentalists), though many businesses are normally supportive of free trade. The gainers are diverse and not well organized, while the lobbies are often well funded and politically connected. Yet world institutions are firmly established to support the causes of freer trade, but this is not so true for the environment.

There is certainly no guarantee that opening trade will increase environmental problems or that closing trade will reduce environmental problems. Yet there is concern that countries with lower environmental standards (or lax enforcement of standards) will have a competitive advantage over countries with higher environmental standards. This has led environmentalists to call for eco-duties to compensate for differing environmental standards among countries. This would mean that a high-standard country could place a tariff on a product imported from a low-standard country to compensate for differing environmental regulations. As will be discussed later, this is not GATT-legal, but environmentalists do not particularly see the GATT as friendly toward environmental causes anyway.

There is no question that the GATT was not formulated with an environmental agenda in mind. It was instituted before the environment became a global issue, and there is no worldwide, all-encompassing environmental agency that can work in parallel with the GATT. When GATT dispute resolution panels are constituted for trade and environment issues, they neither have an environmental representative nor specifically address the environmental aspects of the dispute. The case is judged in its pure economic and commercial form, as will be discussed later.

As will be seen throughout the chapter, the problem with the trade–environment debate is that both sides are correct in many of their observations and tenets, but any solution to the joint problem lies somewhere in the middle of their arguments and arriving at solutions is difficult in many instances. There must be a balancing of environmental benefits with the costs in terms of trade. Cost-benefit analysis is imperative to ensure that environmental standards are not instituted that impose large costs with few benefits. Yet it is clear that the world trading system must pay closer attention to the environment. Are there ways of obtaining the same environmental benefits through measures that are less disruptive to trade?

The solutions must also protect national sovereignty and recognize situational differences among countries. Some developing countries do not have the luxury of stricter pesticide standards because they are concerned with food production and availability. The debate cannot come down to rich countries dictating policies to poorer countries or rich countries limiting access to their markets for poorer countries. There must be more cooperation. Yet some problems, particularly global environmental issues, may be best addressed through increased harmonization of standards that are enacted based on sound science and a cost-benefit evaluation.

EXAMPLES OF CONFLICTS THAT CLARIFY THE DEBATE

Gruenspecht places environmental issues into five categories based mostly on geographical aspects of the problem:

1. Domestic
2. Transboundary

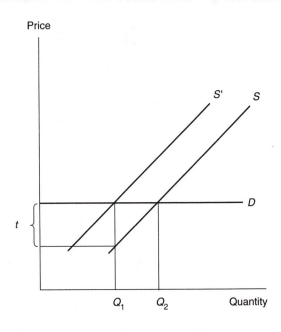

3. Common property rights
4. Offshore
5. Global

Domestic issues are the easiest in some ways because all the differences between private and social costs/benefits are experienced within the country where
the environmental problem takes place. There are many examples of domestic environmental issues: water and air pollution, congestion, pesticide use, depletion of
limited resources. Most economists would argue that taxes or subsidies requiring
the polluter (producer or consumer) to internalize the costs of the externality will
result in an economically efficient solution.

Figure 9.1 shows a simple example where the production of a product generates a negative externality—for instance, the production process pollutes water
that the plant discharges. The demand curve for the product is depicted as D (horizontal assuming that the firm is a price-taker), and the marginal cost of production for the firm is S. The marginal cost curve depicts only the private costs associated with production; it does not include the negative value of polluted water that
comes from the process. When one considers the social costs of the added pollution, the S' curve is obtained. This curve lies above the S curve because of the costs
associated with the polluted water that society must bear. The optimal solution for
the firm is to produce at Q_2 if it does not consider the externality, but a tax of t units
(which forces the firm to internalize the cost of the externality) will encourage the
firm to produce the socially optimal level of Q_1. This is just one of many examples
where a tax system forces the firm to internalize the externality and compels the
firm to produce at the economic optimum.

Transboundary issues occur when pollution is created in one country but the consequences of that pollution are felt in another country. Examples of this are acid rain, which is common in Europe and along the U.S.–Canada border; poor air quality, which often happens along borders; and poor water quality, which is a particularly troublesome issue between the United States and Mexico. These issues are best worked out among the parties involved, and there are rarely trade issues associated with them, unless trade sanctions are used as a mechanism to get one party to the bargaining table.

Common property rights involve resources that are shared among countries because of migratory patterns. Disputes in this area normally involve fishing or harvesting of animals that migrate among country borders. Concerns exist because large harvests by one country will impact other countries that use the common resources. Also, policies that reduce support for migratory animals may reduce their numbers and cause losses (physical or psychological) on other countries where the animals migrate. Fishing rights are a troublesome issue in this category. Again, countries must work together to agree on solutions to problems so that the common resource can be preserved.

Offshore issues are becoming particularly important to environmentalists these days. Such issues deal with endangered species, rain forest preservation, and resource quality in foreign countries. While these issues do not directly influence another country, they do impact the psychological health of citizens worldwide. Americans feel bad when low environmental standards in a country cause individual suffering. They are upset when species become extinct because of poor regulations or enforcement in a country, or when the rain forest is destroyed because of policies that encourage farmers to move every three years. These issues come about because people in one country care about what happens in another country. However, these are difficult issues because they relate to national sovereignty, and the problem often becomes a North–South debate, as mentioned earlier.

Global issues are problems for the entire world whose solution will require actions by a number of countries. The two best examples are global warming and the ozone layer depletion. Only through concerted efforts by many countries will the problems be solved, because the externality is generated by all countries and the accumulated harm is experienced by all. The Montreal Protocol commits signatories to a full phaseout of CFCs by the year 2000 for more developed countries and 2010 for less developed countries. Production of CFCs was stopped in 1996 by most signatories.

GATT IN RELATION TO THE ENVIRONMENT

The overall philosophy of the GATT accord is to encourage countries to participate in mutually beneficial trade. Often, the trade and environment debate focuses on means to force countries into policies or procedures that may not be viewed as beneficial to them. This certainly is the case when the international community wants a country to produce goods using a particular production process. Further, forcing a country to increase its environmental standards might entail trade-offs that the country feels will lower its welfare. Therefore, there is a natural reluctance for the

The Methyl Bromide Case

In most instances, it is estimated that environmental regulations constitute a small percentage of the production costs for most products—1 to 2 percent (Krissoff et al.). However, the effects of banning methyl bromide (MB), a soil and product fumigant, could involve much larger losses on agriculture throughout the world until a suitable substitute is found. The Montreal Protocol calls for regulation of MB use because it affects the ozone layer. In 1994, the U.S. Clean Air Act froze production and imports of MB at 1991 levels, and production and imports are to be phased out by 2001.

The National Agricultural Pesticide Impact Assessment Program of the U.S. Department of Agriculture estimated that a ban on MB would result in economic losses between $1.3 and $1.5 billion for fruits and vegetables. Methyl bromide is used in the United States to fumigate domestically produced fruits and vegetables. It is used on imported fruits and vegetables from many parts of the world, so its banning could also severely restrict fruit and vegetable imports into the United States. Further, many U.S. fruit and vegetable exports rely on MB fumigation to meet quarantine regulations in export markets. The losses come from either increased costs for MB substitutes or losses in consumer and producer surplus as a result of production and consumption changes.

The MB ban is an example of a regulation on a production process that affects trade. Generally, this is not allowed under GATT rules unless it deals specifically with protection of human, animal, or plant life. In this case, since MB has been found to be detrimental to the ozone layer, and the ozone layer protects humans from the sun's ultraviolet light, this ban would likely be upheld under GATT scrutiny. Further, GATT guidelines normally allow international environmental agreements to supersede them, but the role of the GATT is another important issue between environmentalists and free traders. The next section covers GATT principles and rules, and how they relate to the trade and environment debate.

GATT to become embroiled in issues of environmental standards and processes because countries differ in their ideas on what is correct for their own circumstances.

In this context, the main principle of the GATT that applies to the environment is the "national treatment" philosophy: imports should be treated in the same way as domestic production. Countries are free to choose their own policies, but they cannot discriminate against imports. Article 20 of the GATT is not inconsistent with countries establishing their own policies to preserve natural resources and the environment. However, they cannot treat imported and domestic items differently. This would mean that a tax on imported oil would not be a legitimate weapon to combat air pollution from cars. Such a tax is not consistent with national treatment because the tax does not accrue to domestic oil producers. A pure tax on gasoline would be consistent with national treatment because there is no discrimination based on the oil's source.

Article 20 of the GATT accord allows exceptions when countries have trade policies that are "consistent with national environmental policies which are necessary to protect human, animal, or plant life or health." This exception, though, seems to indicate that there is some intrinsic difference between the domestic product and the imported product, or else the domestic product would also be harmful. The "necessary" element of the GATT article is also subject to interpretation. GATT dispute resolution panels have tended to view the term *necessary* to mean that there is no "less GATT inconsistent" or less trade restricting policy available to meet the objective.

This interpretation causes problems with some environmentalists because there is always a less trade restrictive policy that could meet the same objectives, but it might be very difficult or impossible to use. For instance, in the tuna–dolphin case, free traders argue that labeling tuna as "dolphin safe" is less trade restricting and will meet the objectives of killing fewer dolphins. But environmentalists wonder how one could police "dolphin safe" labels.

Article 20 of the GATT accord also allows for trade policies relating to preservation of exhaustible natural resources if these policies are associated with reduced domestic production or consumption. That is, the trade policy must be supported by domestic policies to preserve the natural resource. One cannot have a policy to ban trade in a pesticide if the country produces it.

One could argue that the difference between the domestic and imported product could stem from the production process used, but the GATT does not recognize production process as a legitimate reason for trade barriers, despite the cry of environmentalists. Production processes and methods are strictly domestic matters and are totally determined by the country. However, Esty argues that it is the failure of the GATT to distinguish between product standards (which are GATT-legitimate differences) and production processes (which are not GATT-legitimate differences) that causes the problem. Semiconductors can be produced through a process that uses CFCs. It is the act of using CFCs that is the problem with some semiconductor producers, not any characteristic of the final product. Unless this distinction is allowed, Esty feels that the environmentalists will always be frustrated and the GATT will be less helpful with environmental issues.

It is interesting, though, that the GATT does allow products to be distinguished by the production process for intellectual property (e.g., one cannot simply reproduce a copyright-protected property and export it). However, what constitutes a legitimate or "good" process from a poor one is readily agreed upon by the GATT membership in the case of intellectual property.

Bhagwati argues that there are three key GATT principles relating to the environment. The first is that any GATT member should be able to clarify the true intent of a trade barrier. The GATT dispute resolution mechanism should find out whether the trade barrier or regulation has an environmental or social objective, or whether it is only a means of protecting domestic producers. Second, meeting environmental or social objectives must be done in the least trade disruptive way.

The Tuna–Dolphin Case

The tuna–dolphin dispute between the United States and Mexico is a good case that highlights the concerns of environmentalists, free traders, the GATT, and the use of trade sanctions for an environmental goal. Gruenspecht would classify this as an offshore conflict. The U.S. Marine Mammal Protection Act (MMPA) was passed in 1972 to curtail the incidental killing of mammals by commercial fishermen. The MMPA required the U.S. Secretary of Commerce to take actions to prevent the killing of marine animals by fishermen and to prohibit imports of fish products from offending countries. The main focus of the MMPA was the killing of dolphins when fishermen caught tuna using purse seine nets.

By the late 1980s, U.S. fishing methods had changed, so U.S. fishermen had overcome problems with catching tuna in their nets (and were fishing in areas not frequented by dolphins), but this was not the case with Mexican fishermen, who still killed dolphins while fishing for tuna. An American-based environmental group brought a suit to force the U.S. government to begin following the MMPA, and a U.S. federal judge ruled in the group's favor. The U.S. government decided to ban tuna imports from Mexico, Panama, and Ecuador in September 1990 because of the incidental killing of dolphins (there were certain standards specified by the MMPA that were not being kept by these countries). This dispute was a classic case where the United States decided to enact a trade restriction based on the ideas of its people.

Mexico complained to the GATT because it felt it was being discriminated against. From Mexico's viewpoint, there was no reason that it should change its efficient fishing method because the American public was psycho-

---

Finally, there must be a scientific test of the propositions implicit in the trade barriers. There should be a sound, scientific basis for the alleged connection between the trade barrier or regulation and the desired outcome.

Environmentalists are upset with the GATT because they don't think it upholds any of their principles. Free traders say that the GATT is their institution and it has enough problems within its current mandate. Adding environmental issues to the GATT, especially difficult issues such as differing environmental standards by country, will only make the GATT more controversial and weak.

ECONOMIC PRINCIPLES IN THE TRADE AND ENVIRONMENT DEBATE

There are a number of economic principles that can be brought to the trade and environment debate. Anderson (1992a) provides a good summary. It should be noted from the beginning that most economists would lean toward the free

logically attached to dolphins. Dolphins were not an endangered species and were not covered by any international agreements. A GATT panel ruled that the U.S. import ban on Mexican tuna was illegal because it discriminated against Mexican tuna products. If the United States chose to keep the ban, the panel argued that it should compensate Mexico.

One basis of the GATT panel ruling was that the United States hadn't investigated other means of achieving the objective through less trade restricting policies. The panel felt that the United States was not being consistent in ruling that only tuna from these three countries should be banned. Further, Mexico argued that the ban was discriminating against its tuna based on the production process, which is not a legitimate reason for trade barriers according to the GATT, rather than product standards. Some observers argue that the objective could be achieved by labeling tuna that is obtained in a way that is safe for dolphins (dolphin-safe tuna).

The environmental community was very upset with the GATT panel because it did not consider the legitimacy of the fundamental issue: dolphins were being killed by Mexican fishermen and the United States wanted to do whatever it could to stop it. The ruling seemed to indicate that the United States couldn't protect the health and safety of animals outside its jurisdictional boundaries, which environmentalists thought was too constraining. Many felt that the GATT panel's decision indicated that trade issues were somehow more important or on a "higher plane" than environmental issues. It should be noted, however, that the GATT has never adopted the panel's ruling, so the case has not changed policies at all.

trader viewpoint, though they would realize that environmental degradation entails costs that must be factored into any decision-making rules, regulations, and policies. One of the key questions that must be resolved from an economic perspective is, Where do property rights reside? The generally accepted principle with respect to pollution is that polluters should pay, meaning that property rights (the right of a clean environment) lie with the public.[3] In this instance, the polluter is responsible for paying the damages associated with the externality. This is a basic concept that this section will use, but one must acknowledge that in many instances the damages are very difficult to measure and it is difficult to get the polluter to pay.

---

[3]As noted earlier, poorly defined property rights are often the cause of environmental problems. Chichilnisky argues that these poorly defined property rights often lead to more intensive pollution in less developed countries (LDCs).

FIGURE 9.2
Private and social optimum with a production
externality—no trade.

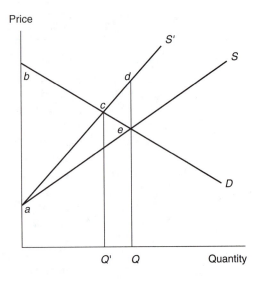

Economics of a Production Externality

When the principle that the polluter pays is adopted, the economic solution to negative externalities is that marginal social benefit will equal marginal social costs because the polluter must incorporate the costs of environmental degradation into production and consumption decisions. The effects of this were shown in Figure 9.1 for a very simple example when there was no trade, the polluting firm was a price-taker, and a tax was imposed to capture the pollution cost. This analysis is extended here to a countrywide situation where firms generate a negative externality as part of the production process. The welfare effects of this situation will be evaluated for no-trade and trade situations.

Assume that the private cost function (supply curve) for the good is S (the sum of all firm marginal cost curves), but the marginal social cost function is S', with the difference being marginal pollution costs (as shown in Figure 9.2). The domestic demand function (reflecting marginal private benefits from consumption) is D. If there is no trade, the optimal market solution considering only private costs is to produce at Q. Social welfare is the area abe (private consumer and producer surplus) minus area ade (pollution costs). The optimal market solution from society's viewpoint comes when the producing firm is forced to internalize pollution costs (through taxes equal to S' minus S) so that firm decisions are made based on S' instead of S. In this situation, production is at Q' and social welfare is area abc. Social welfare is greater by the area cde with taxes because production (and pollution) declines to the socially optimal rate (from Q to Q').

Assume that the country in question is small. When trade is introduced and the country is a net importer of the good (because the world price, P, is below the autarkic price), consumption will increase to C and production will fall to Q_1, if pollution costs are not internalized, or to Q_2, if pollution costs are internalized (Figure 9.3). The country clearly gains from trade liberalization whether or not it is

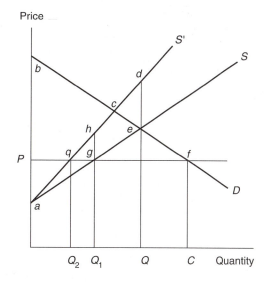

FIGURE 9.3
Private and social optimum with a production
externality— importing country.
*This figure is derived from Figure 2.1 (p. 28)
in Anderson and Blackhurst.*

producing at the social optimum. The gains from liberalizing trade are area *defgh* if the country is producing based on private costs. These gains come about because the country gains area *degh* when it eliminates the losses due to overproduction of the polluting good, while area *efg* is consumer surplus gains (which more than compensate for producer surplus losses) from the increase in consumption levels associated with the lower world price.

If the country is producing based on social costs, the gains from trade will be area *cfq*. All these gains come from gains in consumer surplus, which more than compensate the country for reductions in producer surplus. Notice that the welfare level with trade is higher if the country is producing based on social costs because production is much lower (the country doesn't lose area *hgq* because of pollution). Much of the gain from trade with private costs comes from less pollution, while all the gains with social costs are gains from consumer surplus.

When trade is introduced and the country is a net exporter of the good (because the world price, *P*, is above the autarkic price), the welfare gains are less clear because increased production comes with higher pollution levels. If pollution costs are not internalized, production occurs at Q_1 and consumption occurs at C, with the difference being exported to the world (Figure 9.4). The gain from trade in this case is area *eik* (which is caused by the higher world price that increases producer surplus more than it reduces consumer surplus) minus area *edkm* (which is a loss because production levels are above where they would be if social costs were followed), which could be positive or negative because increased production comes with increased pollution: the net welfare impacts are unclear.

If an optimal tax is imposed on the production process (equal to *S′* minus *S*), the country clearly gains from trade. Consumption remains at *C*, but production falls to the socially optimal rate of Q_2, resulting in trade gains of area *cij* (the only change in welfare comes about because world price increases and the producer

FIGURE 9.4
Private and social optimum with a production externality—exporting country.
This figure is derived from Figure 2.1 (p. 28) in Anderson and Blackhurst.

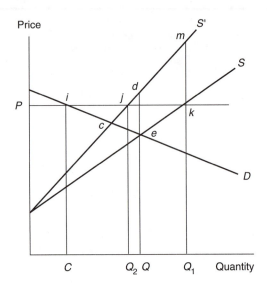

FIGURE 9.5
Export tax versus production tax for an exporting country with a production externality.

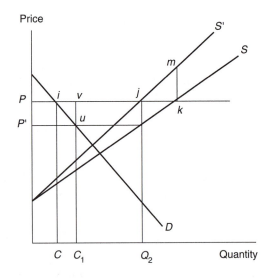

surplus increase is greater than the consumer surplus loss). The difference in welfare positions with the exporting country is that pollution costs are incorporated into the decision-making process in the latter case, so production is lower.

Another policy alternative is to impose an export tax on the product instead of a direct tax on pollution.[4] If an export tax $(P - P')$ is imposed that results in the socially optimal level of production (Q_2), consumption will increase to C_1 because the domestic price for the good falls and production will fall (Figure 9.5). Relative

---

[4]This policy might be enacted because it is much easier to collect an export tax than a pollution tax. Note, however, that export taxes are unconstitutional in the United States.

to free trade when pollution costs are not internalized, the export tax results in a gain of area *jkm* because production (and pollution costs) is lowered to the optimal level, but consumption losses of area *ivu* occur because domestic consumers are consuming more than the socially optimal amount. If area *jkm* is greater than area *ivu*, then there is a welfare gain from the export tax.

The export tax case results in lower welfare because the tax is not paid on all production. The export tax accrues only to international sales, but not domestic sales. Thus, domestic consumers are consuming too much because the price they pay does not include pollution costs. The socially optimal policy is to have the price for all consumers (domestic and international) reflect pollution costs. This result is reflective of many propositions relative to international trade and government policy: it is always best to target the policy to the problem; otherwise, the policy is suboptimal. In this case, the problem is that production causes pollution, so the policy should focus on the production process, not the trade process.

If one moves to the case where the country is large and therefore its production and consumption patterns influence world prices, the key findings above still hold. If the country is an importer, it will gain from liberalized trade because of the normal trade gains, plus it will gain because its production falls and less pollution results. If the country is an exporter, the key finding that a pollution tax on production is best remains, but if such a tax is not imposed, whether there are gains from trade hinges on the difference between private and social costs in the production process. The larger the divergence, the more likely that the exporter will lose from trade liberalization.

Other important cases where trade impacts the environment concern offshore and global issues such as endangered species or ozone depletion. The concepts used above are still valid, but the interpretation of the components changes somewhat. The problem is that the difference between private and social costs for offshore and global issues is often experienced by individuals outside the producing and consuming areas. In such cases, it is difficult to agree on the appropriate tax because the people who are most interested in the tax are neither producers nor consumers: they are outsiders. Further, there is less incentive for the country to enact and collect taxes for the environmental damage when the people who are suffering the loss are foreigners.

The concern over trade in ivory is an excellent example. Most of the benefits from restricting the trade in elephant tusks and other ivory products accrue to Americans and Europeans who are in neither producing countries nor consuming countries. If a ban is not optimal from an economic viewpoint, how could one design a taxation plan that would move the system toward an optimum and how much should the tax be? Certainly this is a difficult if not impossible question to tackle, and its implementation would be horribly complex. The current ban might be optimal when one considers the problems in coming up with an agreement and cost of collecting any taxes.

Economics of International Standards

Production externalities are only one of many environmental conflicts in the trade and environment debate. Consumption externalities are also important, but their analysis is similar to the production externality case. However, an important

Ban on Trade in Ivory

Another example of environmental considerations that are affecting trade is the worldwide ban on the sale of ivory products, which was instituted under the Convention on International Trade in Endangered Species (1973). This ban was established to protect herds of elephants, rhinos, and other animals in Africa that are slaughtered for their ivory, which is demanded by East Asians, who are willing to pay phenomenal prices for it. It is an interesting case where the main proponents of the ban are outside Africa (the producing area) and East Asia (the consuming area). Environmentalists in the United States and Europe have a strong desire to ensure animal numbers in Africa. In order to do that, they have managed to have a ban placed on ivory trade.

Some African nations—particularly Botswana, South Africa, and Zimbabwe— have recently argued vehemently against the ban because they have huge populations of elephants that must be culled (to keep the herds manageable), but are not allowed to sell the ivory. These countries are frustrated because they spend a lot of money on their national parks, despite having rather constrained government budgets, but are denied a large income source (from ivory sales). They agree that there are problems with small herd sizes in some countries, such as Kenya and Uganda, but feel they should not suffer the costs associated with an ivory ban when their herds are beyond the carrying capacity of their lands. This is an example where the marginal benefit of the resource (elephants in this case) differs markedly by location. A harmonized standard of no trade might not be the best policy.

consideration concerning trade and the environment is differing environmental standards by country. If environmental standards differ by country when production externalities are present, then trade patterns will be different than if environmental standards were harmonized. This result bothers many environmentalists, but differing environmental standards do not bother free traders because they are justified from an economic perspective.

Consider a situation where a production externality exists and a country wants to establish an environmental standard that is economically optimal. The standard should be set where the marginal social cost of the damage (MSD) to the environment is equal to the marginal abatement cost of pollution abatement (Pearce and Turner). The marginal social cost curve is positively sloped because as pollution increases, marginal costs of damages are high to society. The marginal abatement cost (MAC) of pollution is negatively sloped because as pollution falls, the extra cost of eliminating it increases (one eliminates pollution that is easier and less costly to remedy at first). The optimal pollution solution is shown in Figure 9.6, where D is the amount of environmental damage. Levels of pollution below D are

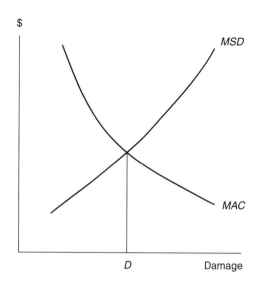

FIGURE 9.6
Optimal environmental standards.

FIGURE 9.7
Losses from harmonized environmental
standards.

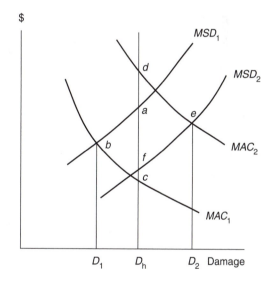

inefficient because abatement costs are too high, while levels of pollution above D are inefficient because damages are too high.

If one views D as the pollution standard, note that it depends on abatement costs and damages, which will vary by location. Because each country has its own unique geographic features, assimilation capacities, cultural traditions, income, and other factors, the MSD and MAC curves will vary by country and the optimal level of pollution will vary by country. Consider a situation, as depicted in Figure 9.7, where the MSD and MAC curves are different between countries 1 and 2 (the subscript identifies the country), but environmental damage is nonetheless harmonized at D_h. The standard in country 1 is too strict, since the optimal environmental damage is D_1, while the standard in country 2 is too lax, since the optimal environmental damage is D_2. The net social welfare loss from this harmonization is area *abc* for country 1 and area *def* for country 2.

As countries develop, the *MSD* curve tends to shift to the left because pollution abatement has a high income elasticity of demand (World Bank). This leads to the conclusion that environmental standards will tend to increase over time as incomes grow, but it also means that there is a natural dispute between the North and the South concerning environmental standards and harmonization (especially if the harmonization is moving toward Northern standards). The North might choose solutions like D_1, whereas the South might choose solutions like D_2. Environmentalists from the North, though, might want the South to have D_1 as the standard.

The framework used in Figures 9.6 and 9.7 can be employed to find the optimal pollution standard. As indicated earlier, environmentalists often favor policies that specify how much pollution is allowed. This comforts them because they know that only so much pollution will enter the environment and they don't feel that companies can buy their way out of pollution. However, these "command and control" systems can be difficult to enforce if pollution sources are dispersed (such as soil erosion) and their discharges cannot be easily measured or monitored. Further, this type of regulation may not provide enough incentive for innovations that will bring about pollution-saving technologies. Market-based taxes and subsidies are better in encouraging firms to innovate and generate advanced technologies that are environmentally friendly because it saves them money.

Effects of Differing Standards

Environmentalists, who are bothered by the idea that pollution and environmental standards should differ by country, are joined in their cry for harmonized standards by more commercial-oriented special interest groups when trade is introduced. The commercial interests in high-standard countries argue that it is not fair for them to compete with companies from low-standard countries. They claim producers in low-standard countries have an unfair advantage.

There is no question that differing environmental standards will convey competitive advantages to producers in some countries over others, as do differing labor standards and antitrust regulations. But before the economics literature on this topic is brought forth, it is useful to open one's eyes to a complete comparison of competitive conditions between countries.

The point has been made many times that less developed countries (LDCs) tend to have lower environmental standards because they are concerned with more pressing needs. Thus, there is no question that everything else being equal, producers of pollution-intensive goods in LDCs will have an advantage over producers in more developed countries (MDCs). However, the assumption that everything else is equal between LDCs and MDCs is very heroic. Visualize yourself as an agricultural producer in Ecuador versus one in the United States. In Ecuador your farm may not have running water, electricity, a sewer system, telephones, access to a highway, or other infrastructure that helps reduce production and marketing costs. You may need to transport your produce by llama to a market that is one day away. As an Ecuadorean farmer, you may not have an extension service for advice, a market news service that reports prices, an experiment station that disseminates research results, a government that subsidizes crop insurance or provides disaster assistance, and the like. For the producer in the United States to complain because Ecuador allows the use of a pesticide that is banned in the United States is short-sighted. There are too many factors that differ between the two situations to make single-item comparisons (such as pesticide regulations) legitimate.

Nonetheless, the fact that environmental standards differ by country and trade in goods is being liberalized (and the world is becoming increasingly integrated economically) raises concerns that these differences will encourage firms to move from one country to another. Copeland and Taylor have shown that with trade liberalization, pollution-intensive firms will have an economic incentive to move from high-standard countries to low-standard countries. The net effect of this movement is that pollution in the high-standard country falls because pollution-intensive firms move out, while pollution in the low-standard country increases.

The reason for the high-standard country's reduction in pollution is because firms are moving to other locations. This might be nice from the environmentalists' viewpoint (unless they are equally concerned by pollution in low-standard countries), but economic development officials are crushed by the job losses from firm exits. The natural tendency is for the high-standard country to do anything possible to keep firms from departing. This might involve tax incentives, job training programs, or reductions in environmental standards; it is the latter that makes environmentalists uneasy. They see a natural tendency for countries to begin a competition based on low environmental standards to save jobs and government incentives. There is not much empirical evidence that standards have been lowered over time by high-standard countries, but Molina has found that there is a tendency for high-pollution (water and solid waste) food processing companies to move to Mexico.

ENVIRONMENTAL CONSEQUENCES OF TRADE LIBERALIZATION IN AGRICULTURE

Some environmentalists were concerned during the Uruguay Round of the GATT negotiations that freer trade in agricultural products would encourage movement of agricultural production to countries with low standards for pesticides, herbicides, and other chemicals, and that the production would take place on fragile soils that are easily eroded. Anderson (1992b) makes the argument that freer trade in agriculture would not only generate large gains from trade but also result in a tremendous reduction in environmental degradation.

Tyers and Anderson estimated the impacts of complete trade liberalization in agriculture. They found that production would move from fertilizer- and pesticide-intensive countries (such as the European Union, Japan, and Korea) to countries that are less intensive users of these land-substituting inputs (such as Argentina, Australia, Brazil, and even the United States). Anderson (1992b) argues that complete liberalization would also result in less intensive feeding of livestock

TABLE 9.1	Fertilizer Use for Selected Countries, Kilograms per Hectare in 1996

Countries That Would Reduce Production with Trade Liberalization

Belgium–Luxembourg	419
Denmark	187
France	260
Germany	233
Ireland	518
Italy	168
Japan	360
Korea	479
Netherlands	548
Switzerland	274
United Kingdom	362

Countries That Would Increase Production with Trade Liberalization

Argentina	31
Australia	38
Brazil	74
Canada	59
New Zealand	202
Poland	113
United States	114

(more pasture systems would be used, as in Argentina, Australia, and New Zealand) and less grain production.

Table 9.1 gives fertilizer use for various countries of the world. The countries with the most protected agricultural sectors (Western European countries, Japan, and Korea) also have the highest level of chemical use.[5] Thus, when trade is liberalized, these countries will produce less agricultural output and there will be a worldwide reduction in fertilizer and pesticide use. Anderson further argues that increasing production in LDCs (which the environmentalists fear) will not have much impact on the environment because the alternative employment for people is a job in an equally polluting manufacturing plant. The alternative employment for farmers in the MDCs is a service job, which results in minimal environmental degradation.

Anderson (1992b) makes the point that domestic policies can have tremendous environmental consequences too. High support prices for grains and other agricultural products are bid into the price of land, making land more expensive relative to chemical inputs. The natural consequence in much of MDC agriculture is to apply large amounts of fertilizers, herbicides, and pesticides to obtain more output per acre. These decisions have environmental consequences, but environmentalists haven't been convinced that trade liberalization in agriculture will improve the environment.

[5]The data on fertilizer use are for all crops, so they do not account for differing fertilizer intensities by crop. Unfortunately, fertilizer application data are not available by crop.

Environmental Aspects of the NAFTA

Environmental issues and their relevance to liberalized trade came to prominence during the NAFTA debate—the first time a free trade agreement was negotiated between an MDC (the United States) and an LDC (Mexico). Environmental problems associated with the maquiladora program became prominent during the NAFTA debate, and there was fear that Mexico would become a pollution haven where companies (and jobs) would flock. The United States and Mexico decided to head off some of these criticisms of possible free trade impacts by negotiating a side agreement on the environment in 1993. This side agreement allowed U.S.–Mexico border issues to be discussed and solutions put forward that would not have happened otherwise. The border—which had suffered from uncontrolled industrial growth, undeveloped environmental infrastructure, lax enforcement of environmental regulations, and the like— had turned into a cesspool (Runge).

The maquiladora program, which began in 1965, allowed companies to have facilities on both sides of the border with duty-free flow of goods between facilities. This was a tremendous boon to both sides of the border. The number of plants along the border has grown at a 20 percent annual rate, generating income increases and many jobs. However, the companies have dumped toxic waste into rivers, spewed air pollution, and presented other environmental problems that made the program controversial.

The 1983 La Paz agreement between the United States and Mexico on Cooperation for the Protection and Improvement of the Environment in the Border Area had many regulations that addressed these problems, but the agreement— which called for construction of waste treatment plants, planning for hazardous waste spills, air emission programs, and so forth—has not been kept. Among the regulations is the requirement that imported toxic chemicals used in maquila production be exported back to the country of origin as toxic waste (Runge). There was a fear that if the La Paz agreement wasn't kept in 1983, little could be done to make sure that environmental aspects of the NAFTA would be upheld.

The Mexican government enacted the General Law for Ecological Equilibrium and Environmental Protection in 1988, which is modeled on U.S. law. It requires companies to file plans and environmental impact assessments for any new construction and plant changes. Another environmental agreement, the Integrated Environmental Plan for the Mexico–U.S. Border Region, was jointly adopted by the United States and Mexico in 1992. These laws and agreements have led to an increase in enforcement of environmental laws: some factories have been closed and others have been forced to reduce emissions by 70 percent. Yet, there was still a concern by environmentalists about NAFTA.

The environmental side agreement for NAFTA was the first time that environmental issues were specifically included in a free trade agreement. The main outcome of the side agreement was the establishment of a Commission on Environmental Cooperation (CEC), which will monitor NAFTA implementation, provide information on compliance with domestic environmental laws, and recommend ways to improve environmental compliance. It will serve as an important sounding board for transborder environmental issues, but it might have a difficult time balancing its oversight and enforcement mandate with independent national policies and organizations.

The CEC should be monitoring and investigating whether trade flows have led to increased environmental degradation. It can also look at such environmental issues as production processes and methods, which are important for agriculture because pesticide residue standards are different between the United States and Mexico (Runge). To date, studies have found that environmental control costs are too low to be a major factor influencing firm movement toward Mexico (Runge). Molina's study in agriculture is the lone exception to this finding.

UNILATERALISM VERSUS MULTILATERALISM

Given the success of the GATT in reducing trade barriers since 1948, it would be nice if environmental issues could be solved through multilateral action so that all countries could agree on the problems and solutions. However, this doesn't seem to happen in most instances. Multilateral environmental agreements that include standards and policies are often difficult to negotiate, and enforcement is also a problem.

Blackhurst and Subramanian made an interesting observation when they said that it took seventy years for countries to negotiate and settle agreements on health and contagious diseases, even though all parties recognized the problems for a long time. With environmental issues, there is a great deal of debate on the causes and cures (as was the case with the health debates), but there is also much debate on whether a problem exists. Suffice it to say that multilateral agreements on the environment are difficult, and the negotiation process is arduous.

Despite the difficulties and time-consuming nature of multilateral agreement on environmental issues, there are successful examples. One of the most important is the Montreal Protocol on Substances That Deplete the Ozone Layer. A major part of this agreement deals with production, use, and trade in chlorofluorocarbons. The agreement calls for restrictions on production and trade among signatories, and also disallows trade with countries that are not signatories to the agreement. This provision is not strictly GATT-legal (because it is not consistent with national treatment), but it persists nonetheless and has not been challenged.

Another example of multilateral action is the United Nations Conference on Environment and Development, which was held in Rio de Janeiro during June 1992. This conference resulted in Agenda 21 and the creation of the UN Commission on Sustainable Development. A number of mandates and agreements have come from these conferences and commissions, which restrict their signatories from trading in certain goods or using certain production processes (Runge).

Yet many environmentalists do not think the multilateral agreements go far enough, and they believe these types of agreements will never go far enough by their very nature. They argue that their provisions are usually noncontroversial and "watered down." The agreements are often plagued by such questions as, How many countries need to sign the agreement to make it "valid," and who decides whether the countries are complying with the agreement? The biggest benefit that a multilateral agreement might convey is that it gives countries an excuse for uni-lateral action (particularly countries from the North). Environmentalists realize that unilateral action can be more swift and targeted because fewer people need to agree on the action. They also feel that progressive unilateral action can lead to successful multilateral action as well.

Multilateral agreements tend to suffer from a lack of participation by many countries. With respect to multilateral environmental agreements, the evidence that a problem exists may not be compelling, so some countries will not want to enter the negotiations. Some countries might accept that a problem exists, but the solution has a low priority because of their social preferences, incomes, or resource endowments. There are simply more pressing issues for them. Another problem is that a country may not agree with the intercountry allocation of responsibilities from the agreement. Often this debate centers on the base year when pollution standards or levels are set, which, of course, changes the required environmental adjustments by signatories. Finally, many countries will simply want to be free rid-ers, where they stay out of the agreement (therefore suffering none of the costs) but receive the benefits of actions from the countries involved. This is why some agree-ments have no-trade clauses for nonsignatories.

LDCs are particularly less likely to become involved in environmental agree-ments. Runge argues that the North should use increased access to its market as a carrot to entice them. Such a commitment would justify improved environmental standards in the South and answer the South's criticism that the North wants the South to sacrifice economic growth so that the North can enjoy a better environ-ment. The South could trade off costs of an improved environment with increased markets in the North.

Multilateral action on environmental issues is important to environmental-ists because they must combat the strong multilateral nature of the GATT. Without some means of arriving at global agreements on environmental standards, en-forcement procedures, and guidelines for trade measures, unilateral measures will be hard to justify. The GATT provides the perfect forum for multilateral agreement on freer trade, but environmental agreements must come together on an issue-by-issue basis. This causes delays and inertia. The strategy of some environmentalists has been to try to destroy or weaken the GATT so that the free trade orientation of the world will be lessened. Esty argues that instead, environmentalists should use the GATT as a model for what he calls a Global Environmental Organization (GEO) that would serve a parallel role to the GATT.

The GEO would exist to provide a framework for countries to fight battles against environmental degradation. It could be a permanent, encompassing orga-nization that focuses on environmental issues among countries, especially com-mon property and border issues, and could even support national environmental issues. It might be similar to the CEC under the NAFTA, but the GEO's role would be broader and involve more open discussion and debate. The GEO could also fund studies that investigate norms, rules, methodologies, and relationships that are important for environmental regulations. If it had sufficient financial backing, the GEO could fund development programs to address some of the LDC environ-mental problems that cannot be addressed because of budget constraints.

Despite whether a GEO is established to promote environmental debates and agreements and to provide a counterbalance (or support) for the GATT, there are special circumstances where environmental agreements must be integrated more formally into the GATT. First, what should be done when trade is restricted based on provisions of multilateral environmental agreements? Do the environmental agreements take precedence over the GATT? The Montreal Protocol prohibits signatories from importing CFCs from nonsignatories. Would this withstand a GATT challenge? Generally multilateral environmental agreements have taken precedence, but there are no cases where this has been disputed.

Other potential problem cases involve unilateral action taken in support of general environmental mandates or agreements. How will GATT dispute panels handle situations where trade restrictions have been imposed on a unilateral basis to support international standards that were composed as part of an international agreement? How will GATT dispute panels rule on trade barriers that are unilaterally imposed to address global or transboundary problems that are officially recognized by international bodies? These are crucial issues that will someday need to be resolved. The ad hoc nature of dealing with the world's environmental problems makes these issues very difficult to handle on a systematic, studied basis.

PRINCIPLES TO RESOLVE THE CONFLICT

Even though direct taxes and subsidies are the optimal economic policies to combat pollution, the political economy of the situation normally results in "command and control" types of restrictions or regulations. Further, there is often significant pressure toward the use of trade barriers to "level the playing field" with respect to environmental standards and production processes. As argued above, this leveling of the field is not efficient from an economic viewpoint because of the vastly different situations each country faces with respect to resource endowments, social preferences, and environmental conditions.

Despite these tendencies for inefficient policy design, there are some principles that can take pressure off the conflict and diminish the likelihood that a confrontation will arise. Runge suggests the following ways for balancing trade and environmental policies:

1. Use trade targets with trade instruments and environmental targets with environmental instruments. It is not good to use an instrument that controls trade (such as an import quota) for a pollution problem. It is better to use a direct tax on pollution. The more targeted environmental policies are on the environment and the more equal the trade burden, the better.

2. Trade policies should aim to reduce trade barriers while remaining environmentally neutral. There should not be a clear trade-off between benefits to trade and costs to the environment that pits one side against the other. Rather, reductions in trade barriers should occur without environmental costs.

3. Environmental policies should be environmentally beneficial while remaining trade neutral. Again, there should be no clear trade-off between benefits to the environment and costs to trade.
4. National governments should be encouraged to pursue similar trade and environmental policy objectives. This does not mean that the policies should be identical, but there should be some common agreement among nations about the types of policies that are to be enacted.

These principles are easily stated, but they are much more difficult to institute because the world is highly integrated these days. It is difficult to keep environmental problems from having trade ramifications and to keep trade liberalization from having environmental ramifications. There are two overriding areas of agreement that can help bridge the gap between environmentalists and free traders. The first is an agreement that the property right lies with a clean environment, so the polluter must pay for pollution. With agreement on this principle, the liability is clear when environmental degradation occurs. The second is the need for collective action and agreements. It would be best if this collective action was accomplished through a GEO or through a significant modification of the GATT, but neither of these is likely. Without these two alternatives, the ad hoc approach of individual international agreements will be the only course. But it is important that the GATT recognize the importance of global environmental agreements, so that the agreements can be used within the GATT framework.

SUMMARY

1. Environmental policies are only one set of domestic policies that have trade consequences. Other nonagricultural policies that have trade consequences include antitrust regulations and labor standards. Because of today's global economy, countries have vested interests in such policies of other countries.
2. There are conflicts between trade and the environment for three reasons. First, trade and trade policies impact the environment because they change production and consumption for countries. Second, environmental policies in one country can impact other countries through trade. Third, people feel that trade policies are a way to force countries into discussions about environmental issues and make them agree to environmental aspects of trade agreements.
3. Environmentalists fear that free trade will bring a downward spiral in environmental standards, but free traders worry that government regulations will get in the way of unfettered competition. There is also a large clash in style between environmentalists and free traders.
4. The most contentious environmental problems deal with common property rights, offshore issues, and global issues. In these instances, many countries must get together for solutions, but they have very different resource endowments and cultural traditions.
5. The GATT has not been friendly to many of the propositions of environmentalists because the GATT's philosophy is to bring countries together for mutually

beneficial trade. Article 20 of the GATT accord allows exceptions when countries have trade policies that are "consistent with national environmental policies which are necessary to protect human, animal, or plant life or health."

6. The economic analysis of environmental problems makes it clear that the best solution is to tackle the environmental externality directly, by taxing the polluter, rather than indirectly through trade policies. Economic analysis also shows that it is best if international standards vary by country depending on marginal social costs of damages and marginal abatement costs.

7. Anderson (1992b) estimates that complete agricultural liberalization would tremendously reduce environmental degradation because countries with high-input agriculture would also tend to reduce agricultural production as a result of liberalization.

8. The NAFTA was the first time a free trade agreement specifically included environmental measures and safeguards.

9. The world might benefit from establishment of a worldwide organization, similar in concept to the GATT, that would implement a mechanism for tackling environmental issues on a global scale. However, it is doubtful that countries would ever agree to establish and properly fund such an organization.

QUESTIONS

1. Are most farmers environmentalists, free traders, or somewhere in between? Explain.

2. Should environmental regulations be standardized so that farmers throughout the world have the same regulations? Defend your answer.

3. Do less developed countries have a good point when they argue that more developed countries have become rich from environmental degradation? What countries are associated with some of the worst pollution stories?

4. Some observers say that the General Agreements on Tariffs and Trade (and the World Trade Organization) are not consistent with environmental interests. Do you agree or disagree? Why?

5. Do you think the United States should invest in the establishment and operation of a Global Environmental Organization?

REFERENCES

Anderson, Kym. "The Standard Welfare Economics of Policies Affecting Trade and the Environment." In Anderson and Blackhurst (Eds.), *The Greening of World Trade Issues.* Ann Arbor: University of Michigan Press, 1992a.

———. "Effects on the Environment and Welfare of Liberalizing World Trade: The Case of Coal and Food." In Anderson and Blackhurst (Eds.), *The Greening of World Trade Issues.* Ann Arbor: University of Michigan Press, 1992b.

——— and Richard Blackhurst. "Trade, the Environment and Public Policy." In Anderson and Blackhurst (Eds.), *The Greening of World Trade Issues.* Ann Arbor: University of Michigan Press, 1992.

Bhagwati, Jagdish. "Trade and the Environment: Exploring the Critical Linkages." In Bredahl, Ballenger, Dunmore, and Roe (Eds.), *Agriculture, Trade, and the Environment: Discovering and Measuring the Critical Linkages.* Boulder, CO: Westview Press, 1996.

Blackhurst, Richard, and Arvind Subramanian. "Promoting Multilateral Cooperation on the Environment." In Anderson and Blackhurst (Eds.), *The Greening of World Trade Issues.* Ann Arbor: University of Michigan Press, 1992.

Chichilnisky, Graciela. "North–South Trade and the Global Environment." *The American Economic Review* 84, No. 4 (September 1994): 851–74.

Copeland, Brian, and Scott Taylor. "Trade and the Environment: A Partial Synthesis." *American Journal of Agricultural Economics* 77 (1995): 765–71.

Esty, Daniel. "Greening the GATT: Trade, Environment, and the Future." Washington, DC: Institute for International Economics, July 1994.

Grossman, Gene, and Anne Krueger. "Environmental Impacts of a North American Free Trade Agreement." *The Mexico-U.S. Free Trade Agreement.* Peter Garber, ed. Cambridge, MA: MIT Press, 1993.

Gruespecht, Howard. "Trade and the Environment: A Tale of Two Paradigms." In Bredahl, Ballenger, Dunmore, and Roe (Eds.), *Agriculture, Trade, and the Environment:Discovering and Measuring the Critical Linkages.* Boulder, CO: Westview Press, 1996.

Krissoff, Barry, Nicole Ballenger, John Dunmore, and Denice Gray. "Exploring Linkages among Agriculture, Trade and the Environment." Agricultural Economic Report No. 738. Economic Research Service, U.S. Department of Agriculture, May 1996.

Lloyd, Peter. "The Problem of Optimal Environmental Policy Choice." In Anderson and Blackhurst (Eds.), *The Greening of World Trade Issues.* Ann Arbor: University of Michigan Press, 1992.

Low, Patrick. "Trade and Environment: What Worries the Developing Countries." *Environmental Law* 23, No. 2 (1993).

Marchant, Mary, and Nicole Ballenger. "The Trade and Environment Debate: Relevant for Southern Agriculture?" *Journal of Agricultural and Applied Economics* 26, No. 1 (July 1994): 108–28.

Molina, D. "Pollution Abatement Costs and U.S.–Mexico Trade in Food Related Products." In G. Williams and T. Grennes (Eds.), *NAFTA and Agriculture: Will the Experiment Work?* International Agricultural Trade Research Consortium and Texas Agricultural Marketing Center, June 1994.

Pearce, David, and Kerry Turner. *Economics of Natural Resources and the Environment.* London: Harvester-Wheatsheaf, 1989.

Rauscher, Michael. "International Economic Integration and the Environment: The Case of Europe." In Anderson and Blackhurst, (Eds.), *The Greening of World Trade Issues.* Ann Arbor: University of Michigan Press, 1992.

Runge, Ford. "Freer Trade, Protected Environment: Balancing Trade Liberalization and Environmental Interests." New York: Council on Foreign Relations Press, 1994.

Tyers, Rodney, and Kym Anderson. *Disarray in World Food Markets: A Quantitative Assessment.* Cambridge, England: Cambridge University Press, 1992.

World Bank. "The World Bank and the Environment." Washington, DC: World Bank, 1992.

Chapter 10

European Agriculture

It is important that students of international trade understand the agricultural situation throughout the world. This is the best way to comprehend world trade patterns and to have a feel for what the future will be like. A full assessment of the world's agriculture is impossible in this book. As a substitute, this chapter covers agriculture in Europe and serves as a model for the data and analysis that would go into assessments of other regions of the world.

European agriculture is chosen because so many of the agricultural policy debates occurring in the United States involve the European Union (E.U.). Without an understanding of E.U. agriculture, as well as its strong and weak points, the student of international trade negotiations will not fully understand the issues. It is also interesting to see how different policies result in a very different agricultural structure.

BACKGROUND ON THE EUROPEAN UNION

As stated in Chapter 7, the European Union is made up of fifteen countries and constitutes the world's largest trading bloc. The E.U. has an area of 3.2 million square kilometers (a bit less than one-third the size of the United States) and a 1996 population of 373 million (41 percent larger than the United States).[1] Its membership is likely to expand markedly in the next decade.

The E.U. is made up of well-developed, high-income countries. Its gross domestic product (GDP) in 1996 was 6.76 billion ecu, or $9.46 billion (the euro wasn't in existence in 1996), which was 15 percent above that of the United States and 46 percent above Japan's. Prices are high in Europe, though, so when the GDP figures are adjusted for purchasing power, U.S. GDP is 2 percent above the E.U.'s.[2] Per

---

[1]Unless otherwise stated, all statistics quoted are from the *Eurostat Yearbook.*
[2]Prices are higher in Japan than in the E.U. or the United States, so Japan's GDP adjusted for purchasing power is only 40 percent of the E.U.'s.

160

capita GDP in the E.U. was $17,323 compared with $25,373 in the United States and $20,250 in Japan. Of the larger countries in the European Union, Germany had the highest per capita GDP ($20,899), followed by France ($19,307), Italy ($19,075), and the Netherlands ($18,990).

Most countries in the E.U. have very low inflation rates (between zero and 3 percent), which is similar to the United States, but unemployment rates are much higher in most E.U. countries. The E.U. unemployment rate in 1996 was 12.5 percent for women and 9.9 percent for men. In 1998, Spain had a very high unemployment rate (approaching 20 percent); Germany, Italy, France, and Belgium also had double-digit rates. The social system and the structure of business taxes on employees are part of the reason for the high unemployment. Unemployment payments to past workers are much higher than in the United States, so there is less incentive for unemployed workers to find jobs. The taxes that businesses pay for employees are higher in Europe, and rules on laying off people are stricter. These policies make firms more reluctant to hire new workers.

Agriculture is less important in the European Union than in the United States. Agriculture accounted for only 2.3 percent of the gross value-added (GVA) in the E.U. economy in 1995, but that percentage varies widely by member country. Agriculture was 14.2 percent of GVA in Greece and 6.0 percent of GVA in Ireland. Because of the population density in the E.U., which is 116 persons per square kilometer versus 28 persons per square kilometer in the United States, agriculture is naturally less important. However, there are pockets in the E.U. where agriculture plays a paramount role.

Contrary to some opinions, the E.U.'s agricultural industry is not made up of many small-scale farmers who are inefficient producers of food. Instead it is a very diverse industry with some very efficient producers, particularly in the Netherlands and parts of France and Germany. Yet the E.U. has struggled with its agricultural policies because it has experienced mounting surpluses, despite constant or falling agricultural prices. It faces the same basic agricultural problem as in the United States: productivity increases outpace the demand for agricultural products.

This growing output from European agriculture was not a problem in 1957 when the precursor to the E.U., the European Economic Community (EEC), was first formed. At that time the six members were net importers of most agricultural products, and the customs union was natural because the surpluses of France, the Netherlands, and Italy could be shipped to Germany, the industrial giant (Italy had a net deficit in some temperate crops, too). The Germans didn't mind supporting agricultural policies that increased farm prices as long as they had better access to export markets for industrial goods. These high farm prices could be supported through an import levy scheme that generated revenue for the EEC (see Chapter 3).

High farm prices had important long-run impacts because farmers throughout the EEC geared up production and the EEC became a net exporter of temperate foodstuffs by 1979. Turning from an importer to an exporter called for a different set of policies because high internal prices continued to be above world market levels. The EEC brought in new members—such as the United Kingdom, Spain, Portugal, and Greece—during the 1970s and 1980s, but those countries increased their agricultural production too because of high EEC prices. The Common Agricultural Policy (CAP) will be discussed further in this chapter.

OVERVIEW OF E.U. AGRICULTURE

Duchene, Szczepanik, and Legg have characterized agriculture in the various countries of the European Union. The following statements about each country are likely oversimplified, but they will give the reader an idea of the background for agriculture in some of the more important countries of the E.U.

France is the leading agricultural exporter in the European Union. It has great resources, particularly in the Paris Basin region in the central part of the country, and the French produce cereals, livestock, and higher-valued crops. There are many efficient producers in France, but there are also many inefficient ones (in the west, southwest, and mountain regions). The French government wants to protect them all. There is no technical advisory service connected to the French government, so Duchene and his colleagues argue that French farmers lack technical expertise and a competitive spirit. The Chambers of Agriculture, which are the extension services in France, are run by commodity groups, and there is little interaction with the French agricultural research infrastructure. The food processing sector in France is also very traditional and has difficulty competing in products with high food safety standards.

The Netherlands is the economic and technical leader of European agriculture. In fact, much of the latest science in livestock and nursery production is applied in the Netherlands. Dutch farmers concentrate on production of high-valued outputs such as livestock (hogs and poultry), dairy, and horticultural crops. Their production facilities are larger in scale than other operations in the E.U., and they are more integrated vertically. The high concentration of livestock has sparked concerns over the environment. There are calls for reduced livestock production and policies aimed at reducing the livestock herds and their pollution.

Germany is a surprisingly important agricultural producer in the E.U., though many of its farmers work only part-time. The country has become an important exporter of agricultural products, particularly processed foods, but it remains a large importer. The Germans' demand for food products has stagnated because of low income elasticities and slow population growth. Yet their general population is concerned with the agricultural economy because they receive noneconomic benefits from the countryside. Germans, and other Europeans as well, love to keep the countryside green and well manicured so that they can enjoy its beauty as they travel outside the cities. They also want agriculture to be friendly to the environment by controlling nitrate runoff and soil erosion.

The United Kingdom and Denmark, which were not original members of the EEC, have a tradition of freer agricultural trade and competition. The average farm size in the U.K. is the largest among E.U. countries, and the U.K. is the most reform-minded as well. The British colonial legacy was built on importing cheap raw materials from foreign countries. Denmark is very export focused and has a highly developed pork industry. Agriculture has less political power in both of these

TABLE 10.1	Value of Agricultural Production in Various Countries of the European Union, 1995	
	Million Ecu	Billion Dollars
France	45,579	$ 63.8
Germany	32,567	45.6
Italy	31,223	43.7
Spain	23,353	32.7
United Kingdom	18,231	25.5
Netherlands	17,112	24.0
Greece	8,497	11.9
Denmark	6,908	9.7
Belgium	6,771	9.5
Total E.U.-15	207,397	290.4

countries relative to others in the E.U., so there is less resistance to changing agricultural prices by the government.

Italy and Spain have the least developed agricultural sectors among the large countries of the European Union. Their agricultural output is more oriented toward Mediterranean crops such as grapes, olives, and vegetables. These countries export these products to the northern E.U. countries, but the products have a relatively low income elasticity of demand and there is substantial competition for them from North Africans. Rural development is a big concern in many areas of Italy and Spain.

E.U. AGRICULTURAL STATISTICS

The E.U. has 139.3 million hectares of agricultural land.[3] Over 76 million of those hectares are arable, with 11.5 million hectares in permanent crops and 50.8 million hectares in permanent grass. France accounts for 23.7 percent of the arable land in the European Union, followed by Spain (19.6 percent), Germany (15.5 percent), and Italy (11.9 percent). Denmark and the Netherlands, two countries that focus on high-valued agricultural products, account for only 3.3 percent and 1.2 percent of the arable land, respectively.

The value of agricultural production totaled 207.4 billion ecu in 1995, which is equivalent to $290.4 billion (U.S. Dept. of Agriculture). France leads the E.U. in value of agricultural production. In 1995 its value reached 45.6 billion ecu compared with Germany's 32.6 billion ecu, Italy's 31.2 billion ecu, Spain's 23.4 billion ecu, and the United Kingdom's 18.2 billion ecu (Table 10.1).

[3]One hectare is equivalent to 2.469 acres.

TABLE 10.2	Value of Selected Agricultural Production for the E.U.-15 in Billion Ecu, 1995
Milk	18.0
Fruits and vegetables	14.9
Cattle	12.5
Hogs	10.8
Cereals	10.2
Other crops	7.6
Poultry and eggs	7.4
Wine	4.7

TABLE 10.3	Land Use in the E.U.-15 in Thousand Hectares, 1995*
Wheat	16,510
Barley	11,024
Olives	4,347
Corn	3,773
Sugar beets	2,129
Oats	2,049
Rye	1,429
Other crops	805

*One hectare is equal to 2.469 acres.

European Union agriculture is split between the north (France, Germany, and the U.K.), where crop and livestock operations dominate, and the south (Italy, Greece, and Spain), where Mediterranean crops dominate. This is reflected in the commodity composition of agricultural output (Table 10.2). The leading commodity in terms of production value is milk at 18.0 billion ecu, followed by fruits and vegetables (14.9 billion ecu), cattle (12.5 billion ecu), hogs (10.8 billion ecu), and cereals (10.2 billion ecu). Table 10.3 shows land use by major crops. Cereals account for much of the land use, but their importance is lower relative to production value. Wheat and barley are commonly used as livestock feed.

Table 10.4 shows total European Union production and the three leading producers. Production of the temperate goods is dominated by the northern countries. Either France or Germany is the leading producer of all major temperate agricultural products. The United Kingdom is also an important producer. Italy is the third leading producer of sugar beets and beef. The dairy industry is very important in the E.U. Pork is the leading meat consumed in Europe, and the E.U. exports a large quantity of it each year.

Yields are high in the E.U. relative to the United States. French, German, British, and E.U.-15 yields per hectare for major crops are presented with U.S. comparisons in Table 10.5. For 1995, the E.U.-15 wheat yield was more than 2.7 times

TABLE 10.4   E.U. Production of Major Agricultural Products in Million Tons

Wheat	100.0	France	36.0	Germany	18.9	U.K.	16.0
Milk	121.6	Germany	28.8	France	25.1	U.K.	14.8
Cereals	206.0	France	62.3	Germany	42.2	U.K.	24.5
Sugar beets	112.0	France	30.7	Germany	26.1	Italy	12.1
Pork	16.3	Germany	3.6	Spain	2.3	France	2.2
Beef	8.0	France	1.7	Germany	1.5	Italy	1.2

TABLE 10.5   E.U. Crop Yields versus U.S. Crop Yields in Tons per Hectare, 1995

	Wheat	Barley	Corn
France	6.53	5.78	7.78
Germany	6.85	5.71	6.71
United Kingdom	7.37	5.58	—
E.U.-15	5.42	4.05	7.80
United States	2.41	3.08	7.12

the U.S. wheat yield, the E.U.-15 barley yield was 30 percent above the U.S. barley yield, and the E.U.-15 corn yield was almost 10 percent above the U.S. corn yield. The wheat and barley yields are very high relative to the United States because the winters in Europe are wet and not so cold, whereas dryland areas in the United States tend to plant wheat and barley because their low moisture levels cannot support other crops.

The major reason for such high crop yields, however, is that land is scarce relative to the United States and land prices are high. The average land value in western Germany for 1995 was over 17,000 ecu per hectare (or $9,640 per acre U.S.); 10,916 ecu per hectare in Italy; 7,608 ecu per hectare in the Netherlands; and 3,080 ecu per hectare in France. These high prices make it crucial for land to be used efficiently, since it is the scarce factor of production. High land prices also result in high fertilizer use, as seen in Chapter 9. Fertilization rates for many E.U. countries are three to five times higher than in the United States.

Land prices in Europe are heavily influenced by urbanization pressures and other nonagricultural uses. The large number of people and the small land mass relative to the United States increase the demand for land. Cash rents for agricultural land are much lower as a percentage of land value than in the United States. In many areas of the E.U., cash rents are only 1 to 3 percent of the land's value, indicating that land owners are obtaining most of their returns on land through capital appreciation. These low cash rents help overcome some of the problems for young people attempting to enter farming, though high land prices make it very difficult to purchase agricultural land.

The E.U.'s Common Agricultural Policy (CAP) is another major reason why land prices are so high. Agricultural output prices have historically been well

	E.U.-15	France	Germany	U.K.
TABLE 10.6 Percentage of Farms in Various Size Categories in the E.U.-15				
Less than 5 hectares	58.4%	27.3%	31.6%	13.8%
5–20 hectares	24.0	21.5	32.5	27.9
20–50 hectares	11.3	24.1	23.3	25.0
50–100 hectares	4.2	17.4	9.1	17.5
More than 100 hectares	2.0	9.6	3.5	15.8

above world levels, and these expected high prices are capitalized into land values. Currently cereal producers in the European Union are guaranteed the intervention price plus a compensating payment. Wheat, corn, and barley have identical intervention (or support) prices of 119.19 ecu/ton[4] (or $167/ton) with another 54.34 ecu/ton ($76/ton) in compensatory payment. When these two rates are combined for wheat, the price to producers is $6.61 per bushel, well above world prices. The intervention prices for milk, beef, and hogs are $19.7/cwt, $221/cwt, and $51/cwt, respectively. All these prices, especially the beef price, are well above U.S. levels. The beef industry in Europe is mostly an offshoot of the dairy industry, so much of the beef is sold as veal calves, hence the high price. Yet the price of beef is still highly distorted relative to world levels. Hog prices are closer to world market levels because the European Union is a major pork exporter. If the internal price is much higher than world prices, export subsidies would be even more taxing on the E.U. budget.

High prices for agricultural products have not only kept people on the farms (there were 16.7 million people living on farms in 1990) but also allowed smaller-scale farms to stay in business. Rural areas are much more vibrant than in the United States because of the large farm populations. This gives rural Europe and its villages a charm that has been lost in rural America. Over one-half of the farms in the E.U. are less than 5 hectares (12.3 acres), while only 2 percent of the farms are over 245 acres. Table 10.6 shows the percentages of farms within size categories for the E.U.-15, France, Germany, and the United Kingdom. Notice that the U.K. has more large-scale farms, but they are still much smaller than typical U.S. farms.

An additional reason for the small scale of most farm operations is the legacy of the eighth-century feudal system (Tracy). At that time, most of the land in continental Europe was controlled by hereditary landlords, who allowed peasants to farm the lands. The rural landscape was slowly transformed over time, and the control of land came into the hands of those who worked it. There has never been a push for farmers to accumulate land because they could make a decent living from relatively small plots.

Napoleonic law, common in continental Europe, also played a role in keeping farms small because inherited land was normally equally shared among heirs.

---

[4]All tons are metric tons, which are approximately 2,205 pounds. The term *cwt* stands for 100 pounds.

	Imports	Exports
Cereals	1,759	4,838
Meat	2,748	3,982
Milk	692	4,617
Fruits and vegetables	12,135	3,967
Oilseeds	4,500	136
Fats and oils	2,640	2,192
Beverages	1,124	8,196
Processed food	601	3,161
Total agriculture	64,167	46,460
Intra-E.U. agricultural trade	119,608	123,369

Over the years, this meant that the average farm size would tend to get smaller as generations passed. Overall, farm size might increase, but this inheritance pattern kept the bias toward the small-scale structure of farms.

These high agricultural prices are passed on to European consumers. The average E.U. resident spends 19.7 percent of household expenditures on food and beverages, about the same as the average Japanese (who spends 19.9 percent), but much more than the average American (who spends 11.4 percent). The percentage spent on food varies from a low of 14.8 percent in the Netherlands to 36.4 percent in Greece. Most of the more developed countries of the E.U. (France, Germany, and the United Kingdom) have averages around 19 percent.

Despite high internal prices, the European Union has been a net exporter of agricultural products since 1979. Table 10.7 shows E.U. exports of agricultural products by subcategory. It is a large net exporter of beverages, such as wine and beer, but it is also a net exporter of meat (particularly pork), cereals, and processed foods. Its main net imports are oilseeds, fruits, and vegetables. In total, the E.U. has a 17.7 billion ecu ($24.8 billion) trade surplus. To support these exports, however, subsidies are needed on cereals, pork, and dairy products.

THE COMMON AGRICULTURAL POLICY, CAP REFORM, AND THE GATT

As stated in Chapter 7, the customs union of Europe began in 1957 with a common agricultural policy. It was significant that agriculture was one common policy area in the early EEC. Agriculture helped solidify the idea of a common Europe, and there was great pride among the agricultural industry that it had helped form a united Europe. The industry has used that as a weapon to stall needed reforms in the Common Agricultural Policy (CAP).

The basic objectives of the CAP were to increase agricultural productivity, ensure a fair standard of living for farmers, stabilize agricultural markets, provide certainty of supply, and ensure that supplies reach consumers at reasonable prices (Tracy). Because of the secular downward trend in agricultural prices throughout the world, the CAP has had the most trouble in ensuring a fair standard of living for its farmers. High agricultural prices have provided enough income for farmers to stay on the land, but their incomes are not comparable to those in nonfarming professions.

The CAP has allowed farm numbers to stay large in the E.U. and has preserved a vibrant rural economy throughout. Villages are economically viable units because there are many farmers in rural areas, which is in contrast to rural areas in the United States. Yet the high prices under the CAP have been capitalized into land prices and made it difficult for new farmers to enter.

Ockenden and Franklin argue that the CAP has not distorted the production of individual crops, but overall agricultural production is likely higher because of the CAP. In other words, the CAP hasn't caused any particular product to be promoted, but all agricultural output is much higher than it would be if prices were at world market levels.

Two large agricultural policy concerns came into the forefront during the late 1980s, and these concerns have continued since that time. The first was about the environment, and this coincided with the political popularity of the Green Party throughout the European Union. Worries about nitrates in drinking water, pastures plowed into cropland, lost wildlife habitat, livestock waste disposal, deteriorated soil structure, and lost hedgerows and wetlands made people throughout the E.U. pay attention to agriculture. They questioned the huge incentives for farmers to produce, especially livestock products.

The second concern was associated with the E.U. budget for agriculture, where expenditures associated with the CAP became an issue in the early 1980s. The E.U. wanted to maintain growth in agriculture and agricultural incomes, but this could not be accomplished without exports, which had budget consequences because high internal prices forced the E.U. to institute subsidies. Agricultural payments (export refunds and price subsidies) were taking over 60 percent of the E.U.'s budget each year (a total of up to $20 billion each year in the late 1980s). Surpluses were mounting, and there was no indication that the situation would improve.

The first reforms of the CAP aimed at "rebalancing" production—moving away from products where surpluses were accumulating rapidly. Cereals, beef, and butter prices were too high, and reform was needed to reduce output incentives. Policy changes began with milk delivery quotas in 1984, restricting the amount of milk that dairy farmers could sell. In 1988, the E.U. imposed "stabilizers" into the CAP, where maximum guaranteed quantities (MGQs) were established (penalties were imposed if production exceeded the MGQ) and set-asides were instituted. The exact policies varied by product, but the idea was to reduce production and therefore reduce CAP budgetary expenditures.

These reforms did little to stem the budgetary pressures on the CAP. Production surpluses continued to mount, and CAP spending skyrocketed. The United States was also putting great pressure on the European Union to change its CAP because the export subsidies and increased production were diminishing U.S. export sales. This time period coincided with the Uruguay Round of the GATT (see Chapter 6) and the oilseed dispute (Explanation Box 10.1), so there was even more

U.S.–E.U. Oilseed Dispute

One of the most bitter disputes between the European Union and the United States involved oilseeds. Soybeans and by-product feed ingredients have entered the E.U. duty-free since the Common Agricultural Policy (CAP) was formed. This has had a large impact on the typical feed ration in the E.U. for many years.[5] It was natural for the E.U. to attempt to switch production from cereals to oilseeds because of the budgetary problems associated with export refunds paid to dispose of cereal surpluses. The E.U. decided to adopt a high internal rapeseed price (rapeseed is the major oilseed produced in Europe) and provide a crushing subsidy to domestic mills in order to compensate them for the high domestic price.

The United States registered a complaint with the GATT saying that this subsidy violated the national treatment provisions of the GATT, since imported oilseeds were not granted a crushing subsidy. When the GATT ruled the subsidies illegal, the E.U. decided to institute direct payments to producers. However, the United States again complained and the GATT again ruled in favor of the United States in March 1992. When the E.U. did not change its policy, the United States announced tariffs on white wine, rapeseed oil, and wheat gluten imports from the E.U. to compensate for lost E.U. markets for oilseeds. An agreement between the United States and the European Union was finally reached in November 1992 to limit E.U. production and to institute set-asides and other provisions that would control E.U. rapeseed production.

call for the E.U. to change its agricultural policies. In June 1991, E.U. Agricultural Commissioner Ray MacSharry developed a comprehensive plan to reform the CAP. It called for reductions in price supports for essentially all temperate agricultural products: grains, oilseeds, beef, and dairy.

After one year of intense debate within the E.U., a modified version of the MacSharry Plan was passed in May 1992 and came to be known as CAP Reform. This reform called for a 33 percent reduction in support prices for cereals over three years and a required 15 percent acreage set-aside to qualify for compensatory payments (which were to compensate producers for the lower support prices). Livestock prices were also reduced with the support price for beef, milk, butter, and milk powder falling by 15 percent, 10 percent, 5 percent, and 5 percent, respectively.

At the same time that the MacSharry Plan was being debated, Arthur Dunkel, the director-general of the GATT, had drafted a plan to serve as the basis of a Uruguay Round agreement on agriculture. This plan, which was presented in

[5]Feed rations commonly include sugar beet pulp, manioc, corn gluten, and citrus pulp with soybean meal used for protein. Tracy states that a typical ration includes 30 units of cereals, 25 units of soybean meal, 17 units of by-products, and 6 units of manioc.

December 1991, was modified by the Blair House Agreement between the United States and the European Union (in November 1992) and served as the basis for the final GATT agreement on agriculture outlined in Chapter 6. The fact that CAP Reform and the GATT agreement happened together was no coincidence. The E.U. had to decide how it wanted to reform the CAP before it could be party to any GATT agreement. Once CAP Reform had passed, it was easy to agree to a Uruguay Round package consistent with the reforms.

The E.U. was forced to bind its tariffs (establish a maximum allowable tariff) as a result of the Uruguay Round GATT agreement. For grains the tariffs were bound at 155 percent of the E.U.'s intervention price, but the actual tariff is calculated by subtracting the import price from an internal reference price, a process similar to the previous variable levy system. However, the actual tariff can never exceed the bound rate. Tariffs on all other agricultural imports were fixed at the bound rates (the levy does not vary within the year).

Because E.U. compensatory payments and U.S. deficiency payments were determined as production-neutral and non-trade-distorting for the GATT, the E.U. has met the Uruguay Round commitments on internal support, and the E.U.'s tariffication process has met the tariff rate reductions required. The European Union could have trouble, however, meeting the required reductions in export subsidies by 2001. Surpluses have continued despite the lower support prices, so the E.U. has continued to subsidize its exports. It is unclear how it will meet its Uruguay Round commitments on export subsidies without drastic measures to curtail output and reduce exports.

There is still intense internal E.U. pressure to reform the CAP. Much of this comes from countries that have applied to be members of the European Union. Poland, Hungary, and the Czech Republic, for example, have large agricultural sectors that could expand greatly if they had access to high E.U. internal prices. Agriculture is the main stumbling block for their entry because of the budgetary problems that might result. Further reform of the CAP is essential if the E.U. is to allow these countries to become part of their union.

SUMMARY

1. Agriculture is less important in the European Union than in the United States; however, there are places in the E.U. where agriculture is very important. This is particularly true in lower-income countries such as Greece and Ireland.
2. Agriculture in the E.U. is quite diverse. France is a major agricultural producer, and the Netherlands has a highly intense, technical agricultural sector. Italy and Spain have the least developed agricultural sectors among the large countries, and their agriculture is more Mediterranean in style.
3. The northern E.U. countries produce large quantities of wheat, barley, milk, and hogs; whereas the southern E.U. countries produce large quantities of wine, olive oil, fruits, and vegetables.
4. Crop yields in the European Union are much higher than in the United States for most products. Land is more scarce, and land prices are higher; output prices are higher because of the CAP, and fertilization rates are also higher.

5. Agricultural production in the European Union occurs on a much smaller scale than in the United States. Most of the farms are smaller than 50 hectares.

6. The E.U. has had to change its CAP often during the past fifteen years because of budgetary expenditures and commodity surpluses. Milk quotas began in the late 1980s, and there were efforts to rebalance crop production in the early 1990s. The CAP Reform package, which was passed in 1992, has been the most comprehensive effort to reduce surpluses through lowering intervention prices and instituting supply control measures.

QUESTIONS

1. How would U.S. farmers change their cropping patterns if wheat was $5.50 per bushel? If land cost $6,000 to $10,000 per acre? Would farmers still have income and cash-flow problems?

2. How will CAP Reform impact rural areas in the European Union? Will their villages be turned into ghost towns, as has happened in many parts of the United States?

3. Agricultural disputes between the United States and the E.U. have been frequent in recent years. How have these conflicts affected the overall economic relationship between the United States and the E.U.?

4. What problems will the E.U. have as it integrates the Central and Eastern European countries into the CAP?

REFERENCES

Duchene, Francois, Edward Szczepanik, and Wilfred Legg. *New Limits on European Agriculture: Politics and the CAP.* Atlantic Institute for International Affairs. Totowa, New Jersey: Rowman and Allanheld, 1985.

European Commission. *Eurostat Yearbook.* Luxembourg: Office for Official Publications of the European Communities, 1997.

European Commission. *The Agricultural Situation in the European Union.* Luxembourg: Office for Official Publications of the European Communities, 1997.

Ockenden, Jonathan, and Michael Franklin. *European Agriculture: Making the CAP Fit the Future.* The Royal Institute of International Affairs. London: Pinter Publishers, 1995.

Tracy, Michael. *Food and Agriculture in a Market Economy: An Introduction to Theory, Practice, and Policy.* Brussels, Belgium: Agricultural Policy Studies, 1993.

U.S. Department of Agriculture. *Agricultural Statistics.* Washington, DC: Government Printing Office.

Chapter 11

Foreign Direct Investment and Processed Food Trade

Exporting is only one of several ways that a company can reach international markets. The most common way is through foreign direct investment (FDI). A firm is understood to have a foreign direct investment when it has an ownership interest of at least 10 percent in a foreign operation. A firm with an ownership presence in more than one country is a multinational enterprise (MNE), and the companies that the MNE owns outside its home country are *affiliates* (foreign subsidiaries) of the MNE. The firm in the home country is often called the *parent*, and the affiliates are located in host countries.

The value of FDI between countries has grown faster than trade since 1985. In that year, FDI flows were $60 billion compared with $315 billion in 1995—a 425 percent increase in eleven years (World Trade Organization, WTO)! It is estimated that the stock of FDI reached $6.1 trillion in 1995, more than merchandise trade for that year, which totaled $4.9 trillion (WTO). These FDI flows come from equity investments through acquisitions or greenfield (new) investments, reinvested earnings from affiliates, and borrowing between parents (the home office) and affiliates. These investments bring not only financial resources but also new technology, managerial and organizational innovations, marketing skills and resources, and other intangible assets. Over time, host countries have been increasingly interested in spillover benefits brought by MNEs that help modernize their industries.

This chapter covers foreign direct investment and the actions of MNEs, including the reasons they invest abroad. Costs and benefits to the home and host countries are also outlined briefly. Finally, the extent of FDI in the food processing, wholesaling, retailing, and food service industries is covered. This important chapter provides the background to understand the globalization of world food industries, a trend that will continue in the future.

FOREIGN DIRECT INVESTMENT CONCEPTS

Not only are the MNEs important investors internationally, but trade among parents and affiliates accounts for approximately 33 percent of international trade. These shipments are classified as intrafirm because there is often no "arms-length" sale between units of the MNE. The prices charged among units (parent and affiliates) of the same MNE are called *transfer prices,* and it is alleged that MNEs often hide profits by using transfer prices that are not reflective of market values. This would allow the MNE to have large profits in low-tax countries and low profits in high-tax countries.

The growth in FDI has been important despite obvious risks and hassles with respect to operating facilities in foreign countries. Exporting is certainly a lower risk strategy for reaching a market because the loss associated with exporting is capped by the value of a shipment. MNEs can lose huge investments if their foreign facilities are nationalized by governments or destroyed through violence or upheaval. It seems logical that a company would find it easier to produce a product in its home country and export it, rather than set up production facilities in foreign lands with different cultures, legal systems, business practices, and the like. Nonetheless, FDI has been and continues to be a very important commercial activity.

Dunning defined three advantages for a firm to invest in foreign production facilities (i.e., to become an MNE): ownership, location, and internalization (OLI). This paradigm is often called the OLI theory of FDI. There have been refinements to this theory since Dunning, but most of the work has been to refine ideas associated with the OLI framework or to integrate the theory of MNEs more fully into the theory of the firm (Ethier, Rugman, Helpman).

LOCATIONAL ADVANTAGES

The location aspects of the framework are easiest to see because they relate to the same factors that cause countries to trade goods. In this sense, FDI (which is in a sense trading of financial and technological inputs) can be viewed as a substitute for trade in goods. Differences in factor costs and endowments—such as lower wage costs, cheaper ingredient prices, higher skilled workers, and other input advantages—are a major reason for a firm to locate an affiliate in a foreign land.

Import barriers can result in a large difference between the landed price for a good (before going through customs) and the internal price of the good. In this case, a company must pay a tariff, obtain a quota, or meet other terms of the trade barrier before the good can enter the potential market. If, instead, the company located a plant in the host country, the affiliate's output would not be subject to the import barrier (the affiliate enjoys "national treatment" in the host country). This is an incentive to locate the plant in another country. Transportation costs also establish a wedge between the price of the product in the exporting country and the importing country, again giving advantages for the firm to locate production in the importing country.

Higher potential growth rates and larger market sizes in host countries are other reasons why a firm might choose to locate a plant in a foreign country. Yet these and other locational advantages alone can be only one part of an FDI theory, because these same locational advantages could be exploited by a domestic firm. A domestic firm could capture the above advantages when it establishes a plant too. Why is the plant established by an MNE, which will have natural disadvantages relative to a domestic firm? The other two parts of the OLI framework provide more insight into why an MNE will invest in the host country.

OWNERSHIP ADVANTAGES

Even if locational differences exist such that certain countries have an advantage over others, this does not explain why a foreign company would find it profitable to invest in the host country. That firm must have some advantages over domestic firms, otherwise domestic firms would exploit the same locational advantages more profitably (since domestic firms would be more comfortable with the business environment, etc.).

Most MNEs are known for having many advantages over the typical domestic firm in the host country. MNEs often bring huge financial resources, technology, patents and other trademarks (brands), and management, marketing and organizational skills that are not available in the host country. The access to new production and information technology, monopolistic power (through patents and trademarks), and marketing skills make the typical MNE a very powerful competitor for domestic firms. Therefore, the MNE can use its ownership advantages to overcome some advantages of domestic firms.

The affiliate of an MNE is also a member of a vast production and marketing network with an organizational structure that reduces costs and enhances the flow of goods among countries. The affiliate can have access to cheaper inputs from the parent or other affiliates, more knowledge of international markets, and much enhanced administrative experience in an increasingly global market. MNEs are often viewed as more flexible because their management is accustomed to working in many different countries. This makes them more likely to change strategies among countries to allow improved operations.

Dunning claims that MNEs tend to be successful in primary goods that require large investments (such as mining), technologically advanced manufacturing (such as automobiles and some electronics), skill or information-intensive industries (such as grain trading), industries where spatial integration is important (such as hotels and airlines), and industries characterized by high trade barriers or transportation costs (such as automobiles). He argues that it is especially important for the MNE to capture advantages that cannot be patented—financial systems, organizational skills, marketing expertise, industrial relations philosophy—because patents expire and can be undermined by product innovations. These nonpatented advantages are more difficult to develop, but they are also more difficult to lose over time.[1]

---

[1] American firms often lead in these nonpatent advantages for many industries. Those advantages are why American negotiators to the GATT and WTO are very interested in liberalizing trade in services (especially banking and insurance), where these nonpatent advantages are vitally important for a firm's success.

Horst claims that food is not a high-technology industry, but FDI in the food industry is still very large and important (as will be discussed later). He thinks that U.S. firms in particular choose to invest overseas because they see the U.S. trends being mimicked by international markets. American firms can take advantage of well-established U.S. trends toward supermarkets, advertising, and product differentiation that are taking hold throughout the rest of the world. Further, American food companies have been diversifying into new products for many years (away from products with slow sales growth), and FDI is a natural part of that diversification process. U.S. food firms are integrating (through FDI) into faster-growing markets that are interested in new products. U.S. food MNEs can perform this efficiently by expanding their distribution and marketing networks because their returns are already large in the United States.

INTERNALIZATION

Even though MNEs may have specific advantages that other firms do not, this does not guarantee that direct investment is the most profitable approach. There are other arrangements that can be devised where the parent company would receive a return on those advantages without being directly involved in investment in the host country. If the arrangements allowed an adequate return on those advantages and saved the MNE from operating in a foreign environment, it seems logical that the parent company would choose to sell its advantages to a local firm in the host country. Such arrangements could involve licensing, consulting contracts, or other standard agreements.

Despite mechanisms that would allow the parent company to receive royalty payments or other income from its partner in the host country, the parent company usually exploits its advantages internally by investing in the host country—thus becoming an MNE. It is important, though, to recognize that the company must receive more profit in internalizing the advantage if it becomes an MNE, otherwise the firm would lose by FDI.

To clarify this point, suppose a company owned a patent on the formulation for a soft drink in the United States and the company wanted to exploit its formulation all over the world. The company could produce the product in the United States and ship it worldwide. However, transportation costs for beverages are quite high relative to value, and it would be much easier for processing facilities in foreign countries to make the product. The company could compete more effectively with locally produced beverages if these cost advantages of locating in other countries were exploited.

The company could build production facilities throughout the world using the special formulation (and hopefully obtaining patent protection in each market where the beverage is produced). Yet it seems much easier for the company to negotiate a licensing agreement with a domestic company in each of the host countries so that a royalty payment is made to the U.S. company for the right to produce the beverage in the host country. If the negotiation process is perfectly competitive, the U.S. company will receive a fair return for use of its formulation without having the hassle and financial burden of operations throughout the world.

Despite the logic of licensing and royalty payments, there are problems with such arrangements between firms, which is why they are much less common than foreign direct investment. One can envision some of the problems from the above example. The U.S. company may be concerned that the local company will not be adept at maintaining product quality or extending marketing activities for the product. The full market potential might not be realized for the product, and the U.S. company would then lose potential profits (particularly since many foreign companies do not have the marketing expertise of American companies). More important, the U.S. company may fear that its partner in the host country will copy the formulation and stop making royalty payments. If this occurs, there will be a legal battle that will be fought in the host country, where the local firm has an advantage over the U.S. firm.

The U.S. company may find it impossible to negotiate a "fair" agreement where it is compensated for its formulation. Markets are imperfect, and there is a lack of information in many markets. This might cause the negotiated royalty payments to be small relative to the return the U.S. company foresees in exploiting the advantage internally. Finally, the U.S. company may incur increased transaction costs in dealing with a separate firm in each host country rather than having its own affiliate in the country. If the parent is dealing with an affiliate, more information can be shared to improve operations in the home and host countries, a common marketing strategy can be adopted, and needed adjustments can be handled more smoothly.

Dunning argues that the costs of exploiting most ownership advantages are often near zero in host countries, so it doesn't take much of an increase in transaction costs or risk in technological or patent seepage to encourage the MNE to invest in a country, rather than sell its advantages. There are examples, though, of advantages that are licensed. Two of the most famous U.S. examples are Coca-Cola and McDonald's. Coca-Cola licenses its patent rights throughout the world, but it protects its formulation by producing the syrup exclusively in company-owned facilities. McDonald's sells franchises throughout the world, but ensures quality through contract specifications. Generally, the more complex the MNE's advantage, the more likely it will exploit the advantage internally (Dunning).

Internalizing the firm's advantages might also result in lower costs or help it compete in the home market. The parent and affiliates can share in certain production and distribution costs, such as research and development, advertising, and so forth. These cost savings are derived from economies of scope, where the firm has lower costs per unit by taking on these extra activities. There are many examples of firms with broad product lines and distribution networks that have been aggregated to lower costs. FDI is simply another way of expanding the firm to take advantage of these lower costs. Even if the MNE's costs are not lower from FDI, some argue that the MNE can use profits from overseas operations to combat its rivals in the home market.

One final word about FDI involves the dynamic nature of globalization. It is not unusual for a firm to reach international markets initially through exporting, because the risks are lower and the fixed costs of exporting are low relative to direct investment. However, as international sales increase, the firm incurs lower cost by taking the risk of the large fixed costs, but lower variable costs, of direct investment (Rugman). In this sense, FDI is the maturation of a firm's internationalization process.

COSTS AND BENEFITS OF FDI

A globally integrated capital market is the result of the world realizing that investment flows (in portfolio and equity form) among countries are beneficial to all. Increasingly sophisticated companies in an information- and technology-intensive world will exploit their advantages whether or not governments desire it. In the last decade, more governments have realized that many regulations on FDI flows only serve to isolate the country from the rest of the world and therefore slow technological diffusion into the host country. Simply put, the home governments realize that they cannot easily keep their MNEs from investing abroad and host governments realize that it is in their best interest to allow MNEs to enter.

The benefits from FDI to the host country are easier to see than the benefits to the home (sending) country. When an MNE decides to build a plant in the host country, there are immediate employment and income gains from the production process. These gains will have direct and indirect benefits that filter throughout the economy. More important, though, is the new technology, management, and competition that the MNE brings. Some of these benefits will spill over to other companies in the host country and improve the management and operations of those companies. There will often be an increased demand for skilled workers at the new facility, and the on-the-job training that comes from the new plant's employment practices will significantly increase the job skills of the existing labor force.

Many potential host countries for FDI had discouraged MNE entry in the 1960s and 1970s because they felt that the multinational enterprises were "buying up assets" or "obtaining undue political influence" through their activities. Entry of a foreign firm was viewed as a loss of national sovereignty, and it hurt national pride. In the late 1980s and early 1990s, many countries, particularly in Asia and Latin America, realized that this attitude was keeping new ideas and technologies out of their country, promoting an inward-looking industrial structure that was not very efficient in delivering modern goods and services.[2] The domestic market structure in these countries was not very competitive, and the countries decided that FDI could immediately force their domestic companies to become more efficient and cognizant of consumer desires.

The United States is a classic example of the benefits of hosting FDI. There was a great controversy in the late 1980s because many state governments in the United States provided lavish incentives for foreign MNEs, particularly automobile manufacturing facilities, to locate a plant within the state. The governments might provide infrastructure improvements, tax holidays, and other benefits associated with

---

[2]Mexico was a classic case of this phenomenon, where U.S. investment in particular was highly discouraged through legal requirements and regulatory hassles. More details are provided on these changes for Mexico in Chapter 7.

the FDI. The state of Kentucky, for example, provided $150 million of incentives for the Toyota Manufacturing Company to locate in Georgetown in 1986. This is clear evidence that states view FDI as tremendously beneficial for economic development. Countries often have a less visible bidding process, but it is clear that MNEs in certain industries with large-scale, technology-intensive production facilities often receive incentives to locate in particular countries or regions.

The benefits to the sending country are less clear and probably less pronounced. The parent may export products to the affiliates, so employment and income in the home country could increase. These benefits will naturally be counteracted by any importing of products from affiliates that were previously produced by the parent. It is possible that the MNE will learn things from its affiliates that it can transfer back to the parent operations to make production more efficient. Finally, equity investors in the MNE, which probably are highly represented by citizens of the home country, will find that the returns from their investment will increase if the affiliates are successful, increasing the home country's income.

A key to the benefits of outbound FDI for the home country is the relationship between FDI and international trade. An early conceptual piece on FDI viewed it strictly as a substitute for trade in the sense that firms chose to invest in host countries instead of exporting to them (Mundell). This type of FDI is called *horizontal*: the MNE performs the same type of activities in the host country as in the home country. If that is the case, there are job losses in the home country because of a drop in exports.

A more realistic view of world trade, though, recognizes that trade in intermediate products is very important and that parents can spin off some production to other countries to make their home country activities more internationally competitive. This is what happens in *vertical* FDI, where the host country activities are used as inputs to or outputs for activities in the home country. In this case, FDI could result in increased exports from the home country even before host country income effects are included. Further, MNEs have diversified product lines so that production facilities in a host country for some products could stimulate demand for other product lines produced exclusively in the home country. FDI could be viewed as synergistic in this multiproduct view, resulting in increased home country exports. The relationship between FDI and trade, though, is still a hotly debated issue.

Despite their effects on home country exports, in most countries MNEs are allowed to follow fair business practices outside their home countries without restriction. There is little that the home country can do to discourage its MNEs from investing internationally. FDI is one of the strategies that MNEs can use to help them maximize profits over all of their operations. Competitive forces will drive MNE decisions, and there is little that the home country can do to thwart outbound FDI.

FOREIGN DIRECT INVESTMENT IN FOOD INDUSTRIES

Many of the familiar U.S. food processing companies are MNEs. Six of the ten largest food MNEs in the world were American, as were fourteen of the top twenty-five in 1993 (Table 11.1 lists the top twenty-five food processing companies). Familiar names such as Philip Morris/Kraft Foods, ConAgra, Cargill, PepsiCo, and Coca Cola top the American list. The world's largest food MNE, though, is Nestlé S.A. from Switzerland, with processed food sales of $36.3 billion in 1993.

Activities of U.S. food MNEs are huge and growing. In 1995, there were 764 affiliates of U.S. food MNEs with assets of $99.6 billion, sales of $113.2 billion, and 554,000 employees (Bureau of Economic Analysis). The gross value-added of these affiliates was $25.2 billion, and the sales from these affiliates were approximately five times the value of processed food exports from the United States. These U.S. affiliates imported $3.4 billion from the United States and exported $2.7 billion back. It is well known that most of this FDI is horizontal in nature (they perform the same activity in their parents and affiliates), and the great majority of the output from U.S. affiliates is sold in the host country (Handy and Henderson).

Table 11.2 shows the top fifteen U.S. MNEs by sales of their foreign affiliates in 1994. Coca Cola was the largest U.S. food MNE with affiliate sales totaling $11.1 billion in 1994. The other fourteen firms are all well known, and each has affiliate sales totaling at least $1.5 billion. Some of the MNEs are concentrated in a few products (Coca Cola and Kellogg), while others have extensive product lines (Philip Morris and Sara Lee). Some concentrate on rather homogeneous products (Archer Daniels Midland and Chiquita Brands), while others have brands that are recognized throughout the world (Coca Cola and PepsiCo).

Table 11.3 shows the top fifteen U.S. MNE exporters (plus three other MNEs in the top twenty-five) for 1994. Notice that there is great overlap between the affiliate sales list (Table 11.2) and the exporter list (Table 11.3); twelve firms appear on both. However, the exports of these MNEs are usually dwarfed by their affiliate sales. Of the eighteen MNEs that are listed in Tables 11.2 and 11.3, only one, General Mills, has exports exceeding foreign affiliate sales (General Mills had affiliate sales of $186 million in 1994). ConAgra and Anheuser Busch had affiliate sales slightly higher than exports. In the other extreme, eleven of the eighteen had affiliate sales over ten times their exports, with the highest ratios for CPC International (58), Coca Cola (47), and Dole Foods (37).

	Company	Country	Processed Food Sales	Total Sales
1.	Nestlé S.A.	Switzerland	$ 36.3	39.1
2.	Philip Morris/Kraft	USA	33.8	50.6
3.	Unilever	UK/Netherlands	21.6	41.9
4.	ConAgra	USA	18.7	23.5
5.	Cargill	USA	16.7	47.1
6.	PepsiCo	USA	15.7	25.0
7.	Coca Cola	USA	13.9	14.0
8.	Danone S.A.	France	12.3	12.3
9.	Kirin Brewery	Japan	12.1	12.1
10.	IBP, Inc.	USA	11.2	11.7
11.	Mars, Inc.	USA	11.1	12.0
12.	Anheuser-Busch	USA	10.8	11.5
13.	Montedison/Feruzzi/Eridania	Italy	9.9	12.3
14.	Grand Metropolitan	UK	9.9	11.2
15.	Archer Daniels Midland Co.	USA	8.9	11.4
16.	Sara Lee	USA	7.6	15.5
17.	Allied Domecq Plc	UK	7.2	7.2
18.	RJR Nabisco	USA	7.0	15.1
19.	Guinness Plc	UK	7.0	7.0
20.	H.J. Heinz	USA	6.8	7.0
21.	Asahi Breweries	Japan	6.8	6.8
22.	CPC International	USA	6.7	6.7
23.	Dalgety	UK	6.7	6.7
24.	Campbell Soup	USA	6.6	6.6
25.	Bass Plc	UK	6.6	6.6

Source: Henderson, Sheldon, and Pick.

	Company	Sales
1.	Coca Cola Co.	$11,080
2.	Philip Morris/Kraft, Inc.	10,113
3.	PepsiCo, Inc.	6,339
4.	CPC International, Inc.	4,780
5.	H.J. Heinz Co.	3,458
6.	Kellogg Co.	2,721
7.	Archer Daniels Midland Co.	2,666
8.	Sara Lee Corp.	2,344
9.	Campbell Soup	2,120
10.	Dole Foods Co.	2,091
11.	RJR Nabisco	1,970
12.	Quaker Oats Co.	1,926
13.	Chiquita Brands International	1,704
14.	ConAgra, Inc.	1,674
15.	Ralston Purina	1,500

Source: Henderson, Handy, and Neff.

TABLE 11.3 Exports (in Million Dollars) of U.S. Food MNEs and Their Ratio of Affiliate Sales to Exports, 1994

	Company	Exports	Affiliate Sales over Exports
1.	Philip Morris/Kraft, Inc.	$1,864	5
2.	Con Agra, Inc.	1,636	1
3.	Archer Daniels Midland Co.	920	3
4.	Anheuser-Busch	732	1
5.	PepsiCo, Inc.	451	14
6.	General Mills, Inc.	305	0.6
7.	Hershey Food Corp.	238	2
8.	Coca Cola Co.	235	47
9.	H.J. Heinz Co.	209	17
10.	Sara Lee Corp.	184	13
11.	Quaker Oats Co.	181	11
12.	Campbell Soup	174	12
13.	Ralston Purina	148	9
14.	Kellogg Co.	131	21
15.	Chiquita Brands International	113	15
(18.)	RJR Nabisco	97	20
(19.)	CPC International	83	58
(21.)	Dole Foods Co.	56	37

Source: Henderson, Handy, and Neff.

Most affiliates of U.S. food MNEs are located in Europe, accounting for 56 percent of the gross value-added; affiliates in Latin America account for 20 percent, affiliates in Asia account for 14 percent, and affiliates in Canada account for 8 percent (Bureau of Economic Analysis). U.S. MNEs choose countries with factor endowments similar to the United States for their affiliates, rather than locating in lower-income countries where factor costs, especially labor, are lower. Research has shown that host country market size, tax policies, and exchange rate differentials are important determinants for the location of U.S. FDI in the food industry. Wages in the host country are not significant determinants of U.S. FDI patterns (Ning and Reed).

Foreign MNEs are also quite active in the United States. In 1995, there were 2,896 U.S. affiliates of foreign food MNEs with assets of $57.2 billion, sales of $50.9 billion, and 228,000 employees. The gross value-added of these affiliates was $12.2 billion, and the sales from these affiliates were also many times the value of processed food imports into the United States. These affiliates of foreign MNEs exported $2.8 billion from the United States and imported $3.2 billion into the United States. European firms are very important investors in the U.S. food industry, accounting for 67 percent of the gross value-added, followed by Canadian and Asian firms.

There has been some analysis of the conditions under which food firms become MNEs and what factors determine their size and success. Reed and Ning, in their analysis of thirty-four U.S. food processing companies, found that the ones that became multinationals were more capital-intensive firms that had extensive, diverse product lines. Moving into the multinational arena was a means to exploit those advantages. Once the firms became multinational, these same factors positively influenced their

MNEs are Important in Trade Negotiations Too

A trade dispute pitting the United States, Ecuador, Guatemala, Honduras, and Mexico against the European Union demonstrates the importance and power of U.S. MNEs. In 1993, the E.U. adopted policies that allowed bananas imported from past English and French colonies in the Caribbean to have greater access and a lower tariff than bananas imported from Central and South America. This sharply limits access of these bananas to an important market. The five countries above filed a petition with the WTO for the E.U. to change its banana import regime—not to equalize the preferences among Caribbean and Central/South American countries but at least to lower the distortions.

One might wonder why the United States is involved in this case, since it exports no bananas and is a large importer. The link is that two U.S. multinational companies, Dole Foods and Chiquita Brands, have huge investments in the Central and South American banana industries. Their business is hurt because it is difficult to export bananas to the E.U. as a result of those preferences.

The WTO agrees with the position of the United States and the other countries. It has ruled against the E.U. three times, and the E.U. continues to stall through legal maneuvers to delay further liberalization. In December 1998, the United States requested the right to withdraw $520 million in trade preferences (or increase tariffs to generate tariff revenue of $520 million) on various E.U. goods if they did not follow the WTO ruling. The panel came back and said that the United States could withdraw $191 million in trade preferences if the E.U. did not liberalize further. At this writing, the preferences have not been withdrawn, but it is clear that U.S. trade policy is meant to preserve market access for U.S. exporters and MNEs.

size. The study also found that MNEs that spent a higher percentage of their budget on marketing tended to be larger, while MNEs that spend a higher percentage of their budget on research and development tended to be smaller. These two latter results would indicate that MNEs have strong brands that they would like to exploit, but they do not tend to have large R&D budgets that are developing new products where ideas might leak to other countries.

Henderson, Voros, and Hirschberg used a 144-firm sample of food companies from throughout the world in their analysis of MNE behavior. They found that smaller firms tend to export, while larger firms tend to invest to reach foreign markets; more specialized firms tend to export, while more diversified firms tend to invest; and firms with a large domestic market share are more likely to invest than export. They concluded that firms were investing abroad to exploit brands and corporate goodwill, especially non-U.S. firms, and that non-U.S. MNEs tended to export a higher percentage of their production than U.S. MNEs.

There has also been rapid growth of FDI in food distribution and food service for the United States (both outbound and inbound investment). In 1993, affiliates of

TABLE 11.4   Sales (in Billion Dollars) and Number of International Establishments for Important American-Owned Restaurant Chains*

Chain	Total Sales	Non-U.S. Sales	Number of International Establishments
McDonald's	$26.0	$11.1	5,461
KFC	7.1	3.6	4,258
Pizza Hut	6.9	1.9	2,928
Dairy Queen	5.5	0.3	628
Taco Bell	4.3	0.1	162
Wendy's	4.3	0.4	413
Subway	2.5	0.3	944
Domino's Pizza	2.5	0.4	840
Little Caesar	2.0	0.1	155
Arby's	1.8	0.1	168

*Includes sales from franchisee-owned establishments.
Source: Restaurant Business and Company Reports.

foreign wholesalers had sales of $21.7 billion in the United States, whereas U.S. wholesalers had sales from their foreign affiliates totaling $15.8 billion (Henderson, Handy, and Neff). Foreign food MNEs have been particularly aggressive in the U.S. food retailing business. Supermarket chains such as Albertsons (Germany)—which is the fourth-largest food retailer in the United States—Atlantic and Pacific Tea (Germany), Food Lion (Belgium), and Ahold (Netherlands) have seen high sales growth in the United States. Total sales of foreign food retailers in the United States reached $51.5 billion in 1993, versus foreign sales for U.S. food retailers of $11.9 billion.

American-owned restaurant chains have spread throughout the world and are an important area of FDI for the U.S. food industry. Sales figures on most restaurant chains include company-owned firms and franchisee-owned establishments, but this is indicative of the MNE's international presence. In 1994, McDonald's was the world's leading restaurant chain with sales of $26 billion, 43 percent of which came from markets outside the United States. Other important U.S.-owned restaurant chains and their sales are presented in Table 11.4. McDonald's, KFC, and Pizza Hut rely heavily on foreign markets.

Foreign-owned restaurant chains are also important in the United States. Well-known names include Burger King, which is owned by Grand Metropolitan (U.K.); Hardees and Roy Rogers, which are owned by Imasco Ltd. (Canada); and Dunkin Donuts and Baskin-Robbins, which are owned by Allied-Domecq (U.K.).

SUMMARY

1. A multinational enterprise (MNE) is a firm with an ownership presence in more than one country. The companies that the MNE owns outside its home country are affiliates (foreign subsidiaries) of the MNE.

2. The value of foreign direct investment (FDI) between countries has grown faster than trade since 1985. In that year, FDI flows were $60 billion compared with $315 billion in 1995—a 425 percent increase in eleven years.

3. There are three advantages for a firm to become multinational: ownership, location, and internalization (OLI). Ownership advantages are firm-specific aspects that make the firm able to produce less expensive or superior products. Locational advantages mean that there are input cost or quality advantages available in foreign countries. Internalization advantages are aspects of the firm that make it necessary for it to control the means of production, rather than license or sell its ownership advantages to others.

4. MNEs benefit the host country by providing jobs, technology, management expertise, and increased competition. MNEs benefit the home country by allowing the firm to learn from its affiliates, to concentrate its high-valued activities in the home country, and to provide inputs to the foreign affiliates.

5. Activities of U.S. food MNEs are huge and growing. In 1995, there were 764 affiliates of U.S. food MNEs with assets of $99.6 billion, sales of $113.2 billion, and 554,000 employees. Most of these affiliates are in Europe.

6. Foreign MNEs are also quite active in the United States. In 1995, there were 2,896 U.S. affiliates of foreign food MNEs with assets of $57.2 billion, sales of $50.9 billion, and 228,000 employees.

QUESTIONS

1. What factors might cause U.S. food exports to grow more rapidly in the future than U.S. direct foreign investment in the food industry?

2. MNEs seem to be getting larger through mergers and acquisitions. Are there reasons that this trend might not continue in the future? Why or why not?

3. What are some ownership advantages that a U.S. biotechnology firm might possess? Why might the firm want to exploit them through foreign direct investment?

4. Should there be restrictions on incentives that state and local governments use to attract foreign companies?

5. Why might an MNE argue against the GATT and trade liberalization?

6. Investigate the nationality of some agricultural firms in your area. Why do you think they chose to locate a facility in the United States.?

REFERENCES

Bergsten, Fred, Thomas Horst, and Tim Moran. *American Multinationals and American Interests.* Washington, DC: Brookings Institute, 1978.

Bureau of Economic Analysis. "U.S. Direct Investment Abroad." Washington, DC: U.S. Department of Commerce. Various issues.

———."Foreign Direct Investment in the United States." Washington, DC: U.S. Department of Commerce. Various issues.

Dunning, John. *International Production and the Multinational Enterprise.* London: Allen and Irwin, 1981.

Ethier, Wilfred. "The Multinational Firm." *Quarterly Journal of Economics* 101, No. 4 (1986): 805–33.

Handy, Charles, and Dennis Henderson. "Assessing the Role of Foreign Direct Investment in the Food Manufacturing Industry." In Bredahl, Abbott, and Reed (Eds.), *Competitiveness in International Food Markets.* Boulder, CO: Westview Press, 1994.

Helpman, Elhanan. "A Simple Theory of International Trade with Multinational Corporations." *Journal of Political Economy* 92 (1984): 451–71.

Henderson, Dennis, Charles Handy, and Steven Neff. "Globalization of the Processed Food Market." Agricultural Economic Report Number 742. Economic Research Service, U.S. Department of Agriculture, 1996.

——, Ian Sheldon, and Daniel Pick. "International Commerce in Processed Foods: Patterns and Curiosities." International Agricultural Trade Research Consortium Working Paper #96-3, May 1996.

——, Peter Voros, and Joseph Hirschberg. "Industrial Determinants of International Trade and Foreign Investment by Food and Beverage Manufacturing Firms." In Sheldon and Abbott (Eds.), *Industrial Organization and Trade in the Food Industry.* Boulder, CO: Westview Press, 1996.

Horst, Thomas. *At Home Abroad—A Study of the Domestic and Foreign Operations of the American Food Processing Industry.* Cambridge, MA: Ballinger Publishing, 1974.

Mundell, Robert. "International Trade and Factor Mobility." *American Economic Review* 67 (1957): 321–35.

Ning, Yulin, and Michael Reed. "Locational Determinants of the U.S. Direct Foreign Investment in Food and Kindred Products." *Agribusiness: An International Journal* 11 (1995): 77–85.

Reed, Michael, and Yulin Ning. "Foreign Investment Strategies of U.S. Multinational Food Firms." In Sheldon and Abbott (Eds.), *Industrial Organization and Trade in the Food Industry.*Boulder, CO: Westview Press, 1996.

Rugman, Alan. *Inside the Multinationals: The Economics of Internal Markets.* London: Croon Helm, 1981.

World Trade Organization (WTO). "Trade and Foreign Direct Investment." Press release. October 1996.

Chapter 12

Competitiveness in the Global Food Economy

Classical trade theory and the gains from trade indicate how a country, in aggregate, will be better off with trade. Resources will flow out of industries within a country where producers are relatively less productive and move into industries where producers are relatively more productive. Relative prices and the production possibilities curve are the key to understanding these gains from trade and the movement of resources from one industry to another.

Textbooks on classic trade theory spend much time on what impact trade has on resource returns (particularly wages and returns on capital) and terms of trade.[1] These books spend little time on the firms that underlie this theory, even though some firms succeed and some fail in each industry. This is characteristic of much economic theory: it does a good job of explaining equilibrium but a poor job of explaining how different firms can face the same output, input, and other markets, yet one firm will be profitable while the other firm will not. One might argue that economists assume away much of the problem in firm and industry success because a common assumption is that firms are identical.

This chapter covers the relatively new literature on international competitiveness, which has more of a business-school foundation. It is an attempt to explain why some nations become the home base for very successful industries and others do not. In this analysis, countries themselves are not competitive, but specific industries are competitive within a country. Resources will shift among industries, as classical trade theory indicates, so that there are some competitive industries. This analysis focuses on the characteristics of those competitive industries and how they develop. Competitiveness also attempts to explain the greatly enhanced role of multinational enterprises (MNEs) in international trade and foreign direct invest-

---

[1]See Ethier for discussion of these and other issues in the general theory of international trade.

ment. Porter is the leading author in this area, and agricultural economists have recently attempted to use his concepts in agricultural markets.[2]

THE ROOTS OF COMPETITIVENESS

The theory on gains from trade would imply that most trade would be between countries with significant differences in factor prices (wages, interest rates, land prices, etc.). However, the reality is that most trade is among more developed countries that have similar labor and other endowments. Trade theory, until recently, focused on undifferentiated products that were produced by firms with identical technology and no scale economies. Factor pools were fixed and relatively homogeneous (low-skilled labor was about as detailed as economists would get).

The world is much different than portrayed in the general theory of trade. Technology has given firms the power to overcome scarce resources (this is the concept of induced innovation, which was first observed in agriculture by Hayami and Ruttan). Firms are constantly changing their products to gain new customers. The demand for specialized labor skills is becoming increasingly important, and specialized educational opportunities have allowed labor skills to advance rapidly as employer needs change. MNEs are increasingly important in trade and investment, presumably because of economies of scope. Porter captured these trends in his concept of competitiveness.

The competitiveness paradigm begins with the idea that nations become rich because they experience sustained increases in productivity (for labor, capital, and other input factors). Their residents have highly productive jobs that pay well, their industries make profitable investments that bring handsome returns, and their regulations establish rules that help firms become more focused and productive. All these situations generate high national income. These rich nations have competitive industries and firms that can afford to pay their employees well because they are highly productive employees, producing the best products available. The country has somehow created an environment where its industries create advantages that overcome any hindrances caused by nature (climate, natural resources) or man (regulations, population pressure).

For a country, this means that the development process should be such that there are firms and industries in the nation that are continually expanding their production and exports. These industries will pay their workers a higher wage because they are more productive and will attract investment because the returns are high. Yet this very process will make other industries less competitive within the country because those industries cannot match the factor returns (especially wages) of the industries that are expanding. This process makes economic sense: the only way these firms and industries can expand (since the overall labor pool is relatively fixed) is by shifting resources toward industries that are becoming increasingly productive and shifting resources away from less productive industries to other countries through foreign direct investment or imports. The firms in these

---

[2]Much of the conceptual material presented in the first sections of this chapter comes from Porter.

less productive industries will either go out of business or shift most of their operations to other countries.

An early dynamic theory of trade that can explain this shifting of resources to foreign locations is the product life cycle developed by Vernon. In the first stage, a firm develops a new product (or a new technology to produce an existing product) and sells it in its home market. In the second stage, as information and technology diffuse, the product is manufactured by foreign companies, restricting the market for the original, innovating company. In the third stage, lower-cost international companies make it impossible for the original firm to export the product at all. In the final stage, the original company cannot manufacture the good in the home country because it cannot compete with low-cost imports. During this process, the foreign production facilities, though, can be owned by the original company through foreign direct investment. Thus, the company itself could be manufacturing the product in many countries, but the home country's production would likely be falling after the first stage.

The MNE is pivotal to the competitiveness paradigm because these firms, with operations in more than one country, specialize in producing and distributing differentiated products using highly productive processes. These companies lead the world in production, management, and marketing in their respective industries. They are key in helping increase the income levels of their home country and keeping their home country competitive. There are generally two measures of competitiveness for a firm or industry in this framework: the amount of exports and outbound foreign direct investment.

The nature of competition in the home country is crucial in preparing firms for the international market. Specifically, if an industry has many firms that compete rigorously within the home market, the industry will be more internationally competitive. If firms in an industry face threats of new product introductions, new firm entrants, and much rivalry among current competitors, then they will be better able to face new competition in the future. If the firms have been forced to deal with very discriminating, fastidious buyers of their products who force the firms in the industry to make constant changes to meet customer demands, then the firms are at an advantage. If the firms can purchase their inputs in markets that are characterized by fierce competition among suppliers where there is constant innovation in inputs, the industry is at an advantage. All these factors may make it difficult for individual firms to survive, but they force them into a mentality such that they are accustomed to such behavior among suppliers, buyers, and competitors that forces firms to constantly consider new ways of doing business.

Porter argues that firms can compete and succeed on two bases. First, the firm can compete by supplying the lowest-priced product. The market is well established, and the firm's output is not significantly distinguishable from other products. Thus, the firm is competing by designing, producing, and marketing the product in the most efficient manner. As long as this results in lower costs than other competitors, the firm will thrive and be competitive. Ultimately, the firm can survive any price cuts that come because it has lower production costs than any other company.

A low-cost strategy is difficult because new firms can often imitate the production process of the firm and cut their production costs over time. With increased globalization, there are always countries where labor costs are low and labor skills have improved enough to begin production of the product. If the production process is not exceptionally complex, the firm will not only face competition in its home market but increasingly face competition in markets with lower labor costs. A technological treadmill can develop such that cost-reducing technologies are developed continually, but producers do not benefit from the new technologies because output prices adjust quickly. Cochrane argued long ago that this is the case in U.S. agriculture. As one can imagine, staying the low-cost producer is no easy task and requires constant investment in cost-reducing technology.

The second way to compete is on the basis of a differentiated product that caters to specific market segments. In this way, the firm distinguishes its output from that of all other firms, so there is less pressure to constantly cut production costs. Granted, there may be other firms that produce similar products, but those products are not identical, so price competition is lower and the firm has more ways to keep existing customers and encourage new customers. Product differentiation provides more durable advantages, but it normally requires constant upgrading of products, technology, and labor skills. The firm cannot stay with the same product too long or other firms will catch up. This strategy requires constant investment to ensure that the firm's products are differentiated.

As the world has become more integrated and barriers to trade and investment between countries fall, firms no longer are constrained to their home market. They can exploit their strategies globally by exporting or investing to reach international markets. Information systems are exceedingly advanced and low-cost, so firms can reach these markets with increasing ease. Depending on economies of scale in production, economies of scope in marketing, transportation costs, trade barriers, differences in local needs, and other factors, firms can decide between keeping production in their home country and exporting to others, or investing in production facilities in another country.

These firm-level decisions have important consequences for the home country and its economic well-being. If countries understand what makes these firms highly competitive, then they formulate policies, establish regulations, and make investments that will lead to more highly productive workers, higher national income, and increased exports and outbound foreign direct investment. At a minimum, governments must help create an environment for their competitive firms and industries to succeed over time. Otherwise, many countries face the prospect of stagnant or declining industries and lower national income. As will be seen, having vibrant, competitive industries does not mean that there are no problems to be overcome, but the resulting industrial structure is much better than the alternative.

PORTER'S FOUR DIAMONDS

How can the home country provide an environment that stimulates its firms to become internationally competitive companies that constantly upgrade their products and skills so that they generate high income levels and returns on investment?

Obviously, the home country must have markets that are dynamic, challenging, and discriminating, so that its firms are accustomed to an environment that forces them to either upgrade or lose market share. The home country must focus on products that have all the ingredients of success for an industry and the dynamics to sustain that success. This doesn't sound easy, but one can imagine that these factors are needed to meet the demands of the ever-changing, fussy consumer.

Porter classified the environment for success into four areas: factor conditions; demand conditions; related and supporting industries; and firm strategy, structure, and rivalry. Each of these areas has aspects that are relevant for a particular industry in a particular home country. Sometimes one area is more important than another, but usually successful firms and industries have positive aspects associated with all four areas. Each area is discussed separately.

Factor Conditions

The basis for the general theory of comparative advantage and gains from trade is the existence of different factor pools among countries. Countries that are land-abundant export land-intensive products, while countries that are labor-abundant export labor-intensive products. The competitiveness paradigm recognizes that factor abundance and endowments are important, but the paradigm views these factors in much finer gradations and posits that factors are not simply endowed: they are developed. The development of higher skilled labor, in particular, is a key element of competitiveness.

Porter classifies factors as basic or advanced. Every country has basic factors, but there is little investment that can be made to change them. These factors are often inherited. Among the basic factors are natural resource endowments, climate, location, and the amount of unskilled and semiskilled labor. Certainly there is nothing that a country can do to change its location (though investments can be made to overcome some locational disadvantages) or climate. Unskilled and semi-skilled labor pools can be changed and upgraded, but in many countries, these endowments are determined by birth and death rates. These basic factors are rarely important in competitive advantage because they are not necessary for a specific industry or are widely available.

Advanced factors are more scarce and require large, sustained investments for development. It is these advanced factors that determine striking differences in competitiveness among countries. Advanced factors include the transportation infrastructure, communications system, number of college graduates, number of research scientists, number of engineers and computer scientists, and other highly skilled workers. Notice that many of these advanced factors come from investments in basic factors (often yesterday's advanced factors are today's basic factors). In fact, the advanced factors can be used to overcome hindrances from the basic factors.

These basic and (particularly) advanced factors that reside in a country are crucial to its general economic success. They set the stage for development of successful industries that will generate much value-added and wealth so that firms can pay high wages for their skilled employees. Yet these basic and advanced factors are not generally associated with individual industries. They exist because of government or personal investments, which can then be used by industries to produce goods and services. In a sense, they are generic factors that can then be molded into specific factors that are used in competitive firms and industries.

Another way of categorizing factors is to distinguish between those that are generalized and those that are specialized. Generalized factors can be employed in many industries, yet can be important for success. A well-developed telecommunications system and other infrastructure, a large percentage of college graduates, and a well-functioning financial system are important in making many industries more productive. Yet these factors are accessible and necessary for virtually all industries.

Factors that are molded for specific industries are called *specialized factors*. They include narrowly skilled people, specialized infrastructure, or other specific capabilities that are related to an individual industry. The development of these specialized factors is riskier than other factors because they are associated with specific industries, and those industries can fail despite substantial investment in factors. Often these investments in specific factors are done by the government, by specialized educational firms, by the industry, by the firms themselves, or some combination. The existence of these specialized factors is essential to the development of competitive industries. Their existence also makes resources difficult to move from one industry to another—hence there is entrenched loyalty to existing industries and sometimes great demands for protection as market conditions change.

Competitiveness based on basic and generalized factors can be fleeting because these factors are natural outcomes from the development process. As countries develop, they increase the quantity and quality of their basic and generalized factor pools, allowing them to compete effectively in more advanced industries. This causes severe problems for nations because their industries are continually subjected to competition from countries moving up the development ladder. If given enough time, nations also find ways to overcome their disadvantages in basic and generalized factors through innovations that save labor, reduce the effects of climate, control space utilization, and require fewer natural resources. Because of the ephemeral benefits of these lower-order factors, it is important for the industry to have advanced and specialized factors to give it an edge.

One can view the colleges of agriculture in land-grant universities as one place where specialized factors are developed. The output of B.S., M.S., and Ph.D. degree holders provides firms with highly skilled individuals who help U.S. agriculture remain competitive relative to the agricultural industries in other countries. The research and extension activities from the U.S. Department of Agriculture and the land-grant system also provide improved factors that enhance agricultural competitiveness. These investments in colleges of agriculture come from the individual students, state and federal governments, and private firms.

Demand Conditions

Having the right people and other resources to develop, manufacture, and market products is very important in today's environment, but it is only one piece of the puzzle. Competitive companies must have ways of knowing the present state of consumer wants and needs, but it is also important to be skilled at predicting future consumer desires. Sophisticated buyers that lead the world in demanding highly differentiated products confer dynamic benefits to industries in their home market. They provide a clearer picture of buyers' needs and entice industries to design and produce innovative products that will meet those demands. The firm that comes up with the next consumer craze (whether it be beanie babies, faster computer chips, or some other product) is sure to be quite successful. Often these consumers demand very specialized goods that meet narrow market segments, but those narrow product segments are exactly where many world consumers will be in the years ahead.

An important way for an industry or firm to anticipate future desires is to have consumers who are discriminating and dynamic and provide constant feedback on products. If a company is selling in a market that constantly demands improvement in product quality or new features, then the firms that sell in that market will be forced to meet those consumer demands by constantly changing product features. When those companies sell in international markets, which may have consumers who are less discriminating and therefore firms that are less accustomed to constant product upgrading, these new products will sell very well.

There are many examples where countries have leading industries because of consumer demand patterns. The United States leads the world in the aircraft industry because of its discriminating customer: the U.S. Defense Department. Many innovations for the defense industry have spinoffs for the civilian aviation industry. In addition, Americans have always been in love with air travel, and the distances between many spots in the United States make it natural for it to lead in technologies and infrastructure related to air travel (airplane design, airport design, safety regulations, and competitive regulations).

Japanese consumers are known for their discriminating tastes with respect to electronics and tend to demand features that others cannot imagine. This is partly because Japanese young people, especially women, tend to spend more years at home living with their parents, despite having a well-paying job. Housing rents in Japan are very high, and there is a stronger family orientation in Japan than in the United States or Europe. These young people living at home have money to spend on expensive stereos, cameras, and other electronic equipment, so they demand the latest high-tech features.

Porter argues that the product life for electronics products in Japan is only six months (versus twelve months in other countries), because Japanese consumers are very conscious of the latest gadgets. Thus, Japanese manufacturers must constantly upgrade their electronic equipment to compete with other Japanese companies. These Japanese manufacturers can take the products that are considered outmoded in Japan and sell them in export markets for a longer period of time, making them very difficult competitors in the world electronics market. In fact, the short product cycles in Japan force their firms to export in order to capture research and development expenditures on the new products required for the home market.

Americans seem more sophisticated relative to computers and information networks. The popularity of personal computers in the office and at home, coupled with the Internet, has vaulted the United States into the lead in computer hardware and software. Investments that have been made in development of the Internet and the information superhighway have supported this computer craze. Americans are also leaders in finance because our consumers are accustomed to making purchases on credit; they lead in risk-reducing enterprises such as insurance and futures markets because free enterprise (and the freedom to have financial losses) is such an important part of the American economy. Risks are higher in the United States, so institutions have developed to reduce risk exposure for firms and individuals.

With respect to the food industry, the Europeans were the leaders for many years because of the long tradition of food quality in continental cuisine (wines, sauces, cheeses, breads, etc.). However, in recent years, the Americanization of food consumption patterns (convenience foods with little time for preparation, fast-food consumption, etc.) throughout the world has increased the competitiveness of the U.S. food industry abroad. The popularity of American-style fast-food chains has reached around the globe, and the influence of U.S. food companies has been enhanced through these trends.

Dynamic, discriminating consumers in the home market make it possible for the firm to anticipate what the future will hold for the rest of the world. There is no question that such consumers will give domestic firms an advantage over their international competitors, assuming that they listen to those consumers.

Related and Supporting Industries

Today's consumers demand increasingly complex products that move through numerous production and marketing stages. Often firms find that it is best if some of those production or marketing stages are performed by other companies: firms are less likely to be fully integrated from raw materials to final retail distribution. In such a world, the firm must rely on other industries to support its activities by supplying high-quality inputs and maintaining or enhancing the quality of the firm's output. If the firm allies itself with upstream and downstream industries that are subpar, this will make the firm less competitive. Whatever the relationship structure among firms in the marketing chain, there must be highly developed coordination so that there can be a continual process of innovation and upgrading to maintain the industry's advantages.

The best situation is for the firm to ally itself with the world's best input suppliers (delivering the best products rapidly at a low price). The suppliers should work with the firm to develop processes and products that would improve product quality and distinguish the firm's output from all others. Communication flows between the firm and its suppliers improve the process of product innovations.

In a similar manner, the firm must ally itself with the world's best downstream firms so that its outputs will be demanded as world market conditions change. A close relationship between the firm and its downstream markets will help the firm make necessary improvements in product quality and increase the competitiveness of the entire supply chain. The information flow, which ultimately begins with the consumer, is vitally important to firm success. Again, if the firm is allied with successful downstream companies, it will be much better off. It is difficult to imagine a firm being competitive if it associates with upstream and downstream firms that are not competitive.

Firm Strategy, Structure, and Rivalry

This part of the competitiveness paradigm deals with how firms are created, organized, and managed, and how rivalry among firms in the home country affects competitiveness. The goals, strategies, organization, management, and other firm-level characteristics vary widely among nations, and these variances can give insights into which industries become competitive in a nation. Some industries will match national cultures and tendencies, while others will not.

The competitiveness paradigm, as espoused by Porter, is most controversial relative to this fourth determinant of competitiveness, particularly as it relates to rivalry. Much of the rivalry among firms can go beyond simple price competition. The competitiveness paradigm suggests that fierce domestic competition and rivalry force firms to lower costs (and prices) and upgrade product quality in order to succeed domestically. This constant pressure to improve makes these domestic firms formidable when they enter international markets.

Some opponents of the competitiveness view on rivalry would argue that domestic competition in many countries has fallen over the last decade, particularly in the United States, as a result of mergers and acquisitions, and this consolidation is leading to larger firms that have less pressure from rivals. Some would also argue that there are industries where the consolidations have diminished competition and allowed firms to charge higher prices and capture larger profits. This debate is left to the future literature in competitiveness and competition policy, except to say that it is difficult to determine whether mergers and acquisitions occur in order to cut costs (and therefore make the resulting firm more competitive), to lower the number of rival firms, or to allow cross-subsidization of operations among many conglomerate activities.

There is no question, though, that firm structures and strategies vary across countries, and this can have an impact on firm and industry competitiveness.

ISO 9000

One way that organizations (or firms) can become more competitive is to participate in the International Standards Organization (ISO) 9000 program. The ISO has been developing voluntary standards and guides for many years, but the 9000 program began in 1987. It is a set of generic management system standards that helps an organization manage its processes or activities—something like a best management practices program. Organizations often have written procedures and instructions about how business is performed so that information flows smoothly and resources are utilized effectively. ISO 9000 is part of a process that documents that control system and makes sure that the system is aligned with features desired by customers. The ISO 9000 program tells an organization what the quality assurance system should entail, but not how to do it.

A firm or organization can obtain an ISO 9000 certificate of conformity, which is issued by a person or organization that has been certified by the ISO 9000 member country's regulatory body. The ISO has nothing to do with the certification process; it only sets up the standards. The regulatory body in the member country decides who is competent and what procedures must be followed to audit and certify firms and organizations. The assessment that firms and organizations are following ISO 9000 standards is a matter for the suppliers and their clients, along with the regulatory body in the specific country. An independent quality certification body lends credence, however, and is often valued highly in the marketplace by customers.

The ISO 9000 program has three specific certifications. The ISO 9001 certification is the most encompassing and includes processes from design and development to production, installation, and servicing. The ISO 9002 is the appropriate standard for an organization that does no design and development. The ISO 9003 is the appropriate standard for an organization that only produces products or service, and tests them to make sure they are reliable. The ISO has recently developed standards for environmental management through its ISO 14000 program, which focuses on ways to minimize the harmful environmental effects of organizational activities.

Chinese and Italian firms tend to be small, family-run enterprises that rely on close relationships among people (most people cannot name a major Chinese-run MNE). There is little communication about the firm outside the family, and all decisions are closely held. Communications with partners and cooperating firms in the marketing system are needed more than exacting product specifications. Japanese and Korean firms, however, tend to be huge conglomerates (called *keiretsu* in Japan and *chaebols* in Korea) that are involved in marketing, trade, and finance for nearly every product in the economy. Companies such as Mitsubishi, Hyundai, and Sumitomo are well known throughout the world. Their operations are on a much larger scale, and the financial power of these huge firms (and their clout with the government) makes them formidable competitors.

One important cultural area that influences firms involves attitudes toward risk-taking and failure. Americans are known throughout the world for a generally positive attitude toward risk-taking. Maybe that stems from the immigrant nature and westward movement of the early settlers. Trying and failing is acceptable in the United States, but that is not true in other countries (e.g., Germany and Japan). There are many successful, rich American entrepreneurs who have made their fortune after numerous failures, yet that is not the case in many other societies. American bankruptcy laws allow businesses to fail and the owners can start anew (if they can find someone to provide needed investment capital).

These attitudes and legal devices make it more natural for American firms to enter into and lead the world in industries that involve more risk. It is common for employees of successful firms to quit and start their own business or for business owners (such as American farmers) to take lower returns for their activities because of their desire for independence. That yearning for independence and control leads to American leadership in many technologically intensive areas where new ideas and products are vitally important.

American attitudes on individuality and risk also lead to the development of industries dealing with risk reduction. America is the birthplace for many financial instruments that reduce or spread risk. The United States leads in the banking and insurance areas because institutions have been developed to allow some of the risks faced by American businesses to be shared with others. These institutions make it easier for American businesses to enter risky ventures because there are mechanisms for them to obtain risk capital from others.

Corporate governance and culture can have an important impact on the way firms from different countries operate. Until recently, it was believed that the short-term viewpoint of American investors kept U.S. companies from making long-term gains in market share throughout the world. This short-term view pervades companies because stockholders judge corporate managers every quarter based on earnings. Banks tend to hold a larger share of the stock in German companies, and these banks tend to have a longer-term view of corporate earnings because they do not trade shares often. The Japanese firms have an even longer-term view of the firm's profitability, such that Japanese firms often trade short-term profits for market share.

Recent performance of U.S. companies, however, has changed some views on America's close attention to the quarterly reports of corporate earnings. The long-term view of German and Japanese companies may slow decision making such that poor decisions are not rapidly discovered and corrected. The increasing influence of U.S. institutional investors (mutual funds) in corporate governance makes American managers more accountable for the firm's performance. This accountability, along with lucrative stock option packages, has business management gurus singing the praises of American corporate governance.

Labor–management relations also vary among firms in different countries, giving some industries advantages in some countries over others. The United States is known for a more combative relationship between labor and management, while the Japanese relationship is more associated with a shared vision. Both of these extremes have moved toward the middle over time as societies change. Yet Japanese culture continues to stress the harmony and oneness of being "Japanese," whereas American culture continues to stress the benefits of "doing your own thing."

A final interesting area where countries differ greatly is in the prestige of certain jobs. Nations tend to have competitive firms in areas that are admired. The Germans and the Japanese are very adept at technical jobs that involve exacting engineering requirements. A Japanese economist once cited a study that compared factory worker performance in the United States versus Japan. If a manufacturing process had a minimum specification for an item, the American workers would tend to produce the product very close to the minimum specification. The Japanese worker, in contrast, would consistently produce above the minimum standard. Some people argue that this is because the Japanese, in general, value technical jobs more highly and obtain more enjoyment out of such work. Porter argues that the Americans and British are much more interested in administrative or managerial positions.

The current fad in terms of corporate culture and governance is likely to change over time as the fortunes of various national firms change, but there is little doubt that there will continue to be fundamental differences in these corporations among countries. Many of these differences come from social systems, religious systems, and family relations/structures. As the world integrates, these differences may diminish, but it is unlikely that such differences will disappear over time, and they will continue to be legitimate factors in determining firm and industry success.

THE DYNAMICS OF COMPETITIVE ADVANTAGE AND GOVERNMENT'S ROLE

The dynamism of competitive advantage is extremely important, especially these days when information systems are highly developed and technologies are very portable. Investments must continually be made in factors to keep up with other firms in the industry. Sustained investments, especially in advanced and specialized factors, will keep the firm ahead of the rest of the industry in terms of lower costs or improved products. Industries do not have to have advantages in all four of the diamonds, but having advantages in one or two diamonds normally leads to advantages in other diamonds (because of increased investments, governmental attention, and other natural adjustments in industries). Thus the dynamic relationships among the diamonds are very important.

Induced innovation and its role in overcoming factor shortages are important in a dynamic context. Prices and wages must send signals to firms and workers concerning what factor pools are needed. If factor pools can be increased (i.e., people increasing their skills for specific industries) or technology can be developed to overcome factor shortages, firms will be able to overcome short-term problems over time. Without these signals, firms will tend to rely on the government or other sources to support their industry through subsidies.

Clustering is an important concept in the competitiveness paradigm. Clustering comes about when many industries work together to create specialized conditions that promote international competitiveness. Clusters can be developed horizontally or vertically and can result in many connected relationships. Information sharing is quite important and can lead to tremendously creative innovations. Vertical clusters generate high-quality, specialized products that move through a narrow marketing chain. Horizontal clusters generate highly competitive firms that share activities across the marketing spectrum. In both cases, clusters develop because firms take advantage of positive externalities generated when factor or demand conditions result in highly competitive firms. The spillovers from these highly competitive firms allow agglomeration of other competitive firms in vertical or horizontal industries: clusters help the competitiveness of the industries become more than the summed competitiveness of each industry.

The role of government in improving the competitiveness of industries within their country is solely through their influence on the four determinants. The government must follow policies that help industries take advantage of their strengths and improve those strengths over time. The government must have antitrust regulations that force firms to compete vigorously in markets and enact regulations that encourage industries to upgrade their products and services. The government should assist in developing an environment that will encourage investments in sustainable advantages. Note, however, that most aspects of these determinants are outside the normal realm of government. This could be viewed as a source of frustration or as a source of comfort, depending on the philosophy of the government.

Government policies and regulations can help industries develop competitive advantage if those policies and regulations are anticipatory. There are examples where U.S. and European countries have enacted controversial environmental regulations despite complaints from industry. When those environmental regulations are later enacted by other countries, the U.S. or European firms that have dealt with such policies in the past have an advantage over their international competitors. Anticipatory policies in areas such as food safety, biotechnology, and other agricultural areas could give American agribusinesses an advantage over their competitors in the long run.

Governments need to make sustained investments in areas that will generate basic factors that help the economy in general (primary and secondary education, university education, highways, general infrastructure). Yet it is investment in advanced and specialized factors that will make a difference in many industries. The role of government in the creation of these factors is unclear; should the government choose certain sectors of the economy for such investment? On what basis should the government choose sectors, and what happens if the government finds that it has chosen the wrong industries for these specialized investments? It is clear that government should take more of an enabling role than a pro-active one.

COMPETITIVENESS ANALYSIS IN FOOD AND AGRICULTURE

Reed and Marchant analyzed the U.S. food industry's international competitiveness relative to thirteen other manufacturing industries. They found that the food manufacturing industry had the lowest percentage of output that was exported

(5.4 percent) and had the second-highest ratio of foreign-affiliate sales to U.S. exports (2.33). So U.S. firms were much more likely to process food in their overseas affiliates than to export it from their U.S. plants. Further, most of the investment overseas in food manufacturing facilities was horizontal in nature (where the company is performing the same activities in the host country as it performs in the home country).

Reed and Marchant found that the sixty-four largest U.S. food manufacturing firms had sales from their foreign affiliates that were ten times the size of their U.S. exports, indicating that large firms were much more likely to reach foreign food markets through investment rather than exports (this is consistent with the findings of Handy and Henderson). Further, there was little trade between the foreign affiliate and the U.S. parent. They concluded that if one measures competitiveness in the Porter sense, then the U.S. food manufacturing industry is not very competitive because its exports are low and outbound foreign investment is smaller than in other industries.

Abbott and Bredahl argue that the agricultural industry should be divided into four economies for an analysis of competitiveness: the production of undifferentiated primary commodities, the production of differentiated primary commodities, the conversion of primary commodities into semiprocessed products, and the conversion of primary and semiprocessed products into consumer-ready products. They argue that the United States has historically concentrated on the undifferentiated primary commodities.[3]

Table 12.1 classifies the determinants of competitiveness that are particularly important for agriculture and food. Some of the determinants are endowed, but most are generated by the firms (technology, firm strategy, and product characteristics), individuals (human capital and managerial expertise), clusters of firms (industry structure, input supply, and marketing and distribution channels), or the government (infrastructure and externalities, regulatory environment, and trade policy).

Table 12.2 categorizes the importance of competitiveness determinants among the four economies of agriculture. Natural resource endowments and cost-reducing technologies are crucial for competitiveness in undifferentiated primary commodities. They are less important for differentiated primary commodities, semiprocessed products, and consumer-ready products. Human capital and managerial expertise, quality-enhancing technology, and product characteristics are critical for semiprocessed and consumer-ready products. Firm strategy, the regulatory environment, and trade policies play an important role in all four economies, but their effects can vary widely depending on whether competition is cost-based or product-based.

One critical observation is that product differentiation and marketing chain control are becoming increasingly important in U.S. agriculture as a result of food

---

[3]The percentage of U.S. exports in bulk form has fallen consistently in recent years. In 1990 it was 53 percent, but in 1997 it was only 41 percent (Economic Research Service). However, these percentages are still much higher than the bulk export percentages from the European Union.

TABLE 12.1 Determinants of Competitiveness

Determinant	Competitive Factors
Factor endowments and natural resources	Natural resource advantages are of particular importance to agricultural commodities because soil, climate, and other natural conditions can determine where crops may be grown successfully.
Technology	Country-specific advantages determined by technology serve as the fundamental basis for a competitive advantage. Technical change may be cost reducing and/or quality enhancing.
Investment	The means by which technical change and industrial evolution are accomplished is investment. Market and technical factors affect investment strategies.
Human capital	Human resources are critical to competitiveness of specific products or sectors. Few enterprises require truly unskilled labor. Expertise is critical to an enterprise, but it need not be offered from the same nation that produces the final good sold.
Managerial expertise	Case studies point to firm failures when all signals seem to point to success, because of mistakes made by management.
Product characteristics	Tastes and preferences differ across or within nations, and successful business strategies can be designed around serving market niches in addition to broad market demands. Other nonprice factors such as reliability, maintenance, and service can be important components of product characteristics to both processors and final consumers.
Firm strategy and industry structure	There may be several successful firm strategies. Cost leadership is one option, whereas serving market niches may not demand the lowest production cost.
Input supply	Few enterprises are completely integrated vertically. Relationships between producers and their input suppliers can be critical to the success of a firm or a product.
Marketing and distribution channels	The system to market products and especially to penetrate export markets can be crucial to the success of firms.
Infrastructure and externalities	Governments are responsible for the necessary infrastructure including public works, utility regulation, education, and other public goods. Often the determining factor is external benefits accruing not to a single enterprise but rather broadly to many possible ventures.
Regulatory environment	The government sets the rules of the game under which a firm must proceed, and these can be exceedingly specific and complex, constraining a firm's decisions and opportunities.
Trade policy	Trade policy is a special case of the set of regulations imposed on a firm and relates to products crossing borders. In agriculture, it is the set of domestic policies (or at least domestic objectives), more than the trade policies, that determine the environment within which firms compete. A critical yet unanswered question is when trade policy *per se* is an appropriate strategy for government.

Source: Council on Agricultural Science and Technology.

TABLE 12.2 The Four Economies of Agriculture and the Importance of Determinants of Competitiveness

Determinants of Competitiveness (Value-Added)	Production, Assembly, Transformation (Processing), and Final Distribution of			
	Undifferentiated Primary Commodities (Little Value-Added)	Differentiated Primary Products (Some Value-Added)	Semiprocessed Products (Great Value-Added)	Consumption-Ready Products (Greatest Value-Added)
Natural resource advantage, factor endowments	Generally critical, but the mobility of technology is likely diminishing its importance		Little importance, but varies with mobility of primary inputs	Little importance, but varies with mobility of primary and semi-processed products
Cost-reducing technology	Mandatory, but technology is increasingly mobile	Some importance, but product differentiation requires that characteristics be reflected in production practices; technology generally mobile		
Human capital and managerial expertise	Some importance; skills in application of production technology important; many people involved		Great importance; skills are critical, especially in organization and coordination of activities with fewer people involved	
Quality enhancing technology	Some importance; transportation, for example	Some importance; quality and product form are closely related	End-use characteristics are most important	
Product characteristics and nonprice factors	Grades and standards provide information	Product differentiation possible through quality differences	Cost leadership and product differentiation, or a combination, may be pursued	Cost leadership and product differentiation, or a combination, may be pursued
Firm strategy	Minimum cost is only feasible strategy	Cost and differentiation are possible strategies	Cost leadership and product differentiation, or a combination, may be pursued	Cost leadership and product differentiation, or a combination, may be pursued
Industry structure, input supply, marketing and distribution channels	Markets provide vertical coordination	Depends on economies of scale in economic activities other than production; markets or hierarchies link primary product production; often accomplished by single firms; importance of end-use characteristics at farm level varies and influences vertical coordination of markets		
Infrastructure	Important to cost competitiveness		Important to cost competitiveness, product differentiation, and innovation	
Regulations, environment, and trade policies	Great, but declining; may determine trade patterns	Varies greatly; policies greatly influence competitiveness and trade patterns, but often policy impacts are indirect; technical barriers matter most		

Source: Abbott and Bredahl.

safety concerns and increased consumer demand for high-quality, differentiated products. Highly specialized, identity-preserved products can be supplied easier through biotechnology, and these products are being sold to consumers who are willing to pay more for this quality assurance. These capabilities are giving U.S. food firms an advantage over their foreign competition in many cases.

Bredahl, Abbott, and Reed include a number of country case studies on competitiveness. Each of these studies takes a different approach, but it is interesting to compare their findings nonetheless. MacDonald and Lee investigate the competitiveness of the U.S. food processing industry and conclude that increases in labor productivity in the United States have far outstripped labor productivity advances in other countries. They take a more macro-oriented approach to competitiveness because they do not deal with individual sectors. If the analysis included raw agricultural commodities, U.S. competitiveness would certainly come from its natural resource base, the technology (cost-reducing and product-expanding) generated from the land-grant university system, and its infrastructure (water and land transportation systems).

Miner used many measures of competitiveness in his study of Canada's agri-food system: market share, export orientation, import penetration, labor and capital productivity, product differentiation, technological application, and growth in sales and profits. He used the same competitiveness definition as the Canadian Task Force on Competitiveness in the Agri-Food Industry: "the sustained ability to profitably gain and maintain market share." Most of his findings were in comparison to the United States. He found that U.S. food processing firms (relative to Canadian firms) were larger in terms of value-added and employment, had higher levels of productivity, spent more on research and development, and were more involved in technology transfer. In Canada, agri-food firms were forced to pay more for agricultural raw materials and faced less domestic competition, but were more export-oriented. He believes that Canadian government policies to protect markets and manage suppliers hurt the country's international competitiveness. Canada's most regulated agri-food industries are also Canada's least competitive.

Lattimore used a competitiveness index, which was New Zealand's export share for a commodity divided by New Zealand's share of world production, to measure international competitiveness. He concluded that New Zealand was competitive in apples, beef, mutton, cheese, and butter because the competitiveness index was above 3.0 for all those commodities (indicating that New Zealand's export market share was over three times its world production share). He attributed New Zealand's competitiveness to its natural endowments (weather), large investments in land (improving soil quality) and technology (such as new fruit varieties), and free trade environment (the competition it promotes). New Zealand has a number of commodity marketing boards, and Lattimore was unsure whether these boards (which control international sales) increased or reduced competitiveness.

Walter-Jorgensen discussed the historical development of Denmark's agricultural sector, which exports over 66 percent of its production. This emphasis on exports has forced the Danish agricultural sector to use "focused differentiation." The Danes focus on products with large markets (pork, potted plants, furs), but they differentiate their products from those of other countries through technology and product characteristics. Most of the output of these products comes from large marketing and processing cooperatives that are controlled by farmers, and there is little governmental support (other than export subsidies for pork that compensate producers for high internal feed prices). Denmark has also experienced lower wage cost growth and higher productivity growth than other European countries. Walter-Jorgensen estimates that Danish productivity growth in agriculture was 6 to 7 percent annually in the 1970s and 1980s, versus 4 to 5 percent annual growth in other European Community countries.

All of these studies indicate that countries are very interested in the competitiveness of their agricultural and food processing sectors. Certain elements of those sectors appear competitive, though there are many measures of competitiveness that one can use. What is clear is that economists and policy makers are struggling with strategies that will help their firms compete in an ever-changing, increasingly global market. It is clear that sustained investments in education, infrastructure, and science are needed to maintain and enhance competitiveness, but what other government activities are needed is unclear.

SUMMARY

1. The competitiveness paradigm posits that nations become rich because they experience sustained increases in productivity. Their residents have highly productive jobs that pay well, their industries make profitable investments that bring handsome returns, and their regulations establish rules that help firms become more focused and productive.
2. There are two ways for firms and industries to compete: on a low-cost basis and on a differentiated product basis. There are two measures of competitiveness for a firm and industry: the amount of exports and outbound foreign direct investment.
3. Porter believes that there are four sets of factors that determine the competitiveness of firms and industries: factor conditions; demand conditions; related and supporting industries; and firm strategy, structure, and rivalry.
4. Successful firms have means of anticipating what purchasers want in products, and the firms adapt to supply those products. The rivalry among suppliers is also ruthless so that successful firms are forced to innovate often in terms of cost savings and improved product quality.
5. Clustering of similar industries is often present when a country is competitive. Information sharing, common research and development, and other specialized infrastructure make a difference and can provide advantages to entire industries with linkages among themselves.
6. The U.S. food processing sector is less competitive than other U.S. sectors because of its small percentage of exports; U.S. food processors also have less FDI than other sectors. However, there has been substantial growth in U.S. food exports and investment in recent years.

QUESTIONS

1. In what sense would the closing of a manufacturing plant in your local community be viewed as good news in the competitiveness framework? How is it bad news too?
2. In what ways are American food consumption habits leading demand conditions in the rest of the world?
3. Provide an example where rivalry in the U.S. market has increased agricultural exports. Provide an example where rivalry in the U.S. market has not increased agricultural exports.
4. What are some competitive advantages that a U.S. biotechnology firm might possess?
5. What can your local government do to attract competitive firms in competitive industries?

REFERENCES

Abbott, Philip, and Maury Bredahl. "Competitiveness: Definitions, Useful Concepts, and Issues." In Bredahl, Abbott, and Reed (Eds.), *Competitiveness in International Food Markets.* Boulder, CO: Westview Press, 1994, pp. 11–36.

Bredahl, Maury, Philip Abbott, and Michael Reed. *Competitiveness in International Food Markets.* Boulder, CO: Westview Press, 1994.

Cochrane, Willard. *Farm Prices, Myths, and Reality.* Minneapolis: University of Minnesota Press, 1958.

Council on Agricultural Science and Technology. "Competitiveness of U.S. Agriculture and the Balance of Payments." Task Force Report No. 125. CAST: October 1995.

Economic Research Service. U.S. Department of Agriculture. *Foreign Agricultural Trade of the United States.* Washington, DC: Government Printing Office. Various issues.

Ethier, Wilfred. *Modern International Economics.* New York: W. W. Norton and Company, 1988.

Handy, Charles, and Dennis Henderson. "Assessing the Role of Foreign Direct Investment in the Food Manufacturing Industry." In Bredahl, Abbott, and Reed (Eds.), *Competitiveness in International Food Markets.* Boulder, CO: Westview Press, 1994, pp. 203–30.

Hayami, Yujiro, and Vernon Ruttan. *Agricultural Development: An International Perspective.* Baltimore: The Johns Hopkins Press, 1971.

Lattimore, Ralph. "Assessing the International Competitiveness of the New Zealand Food Sector." In Bredahl, Abbott, and Reed (Eds.), *Competitiveness in International Food Markets.* Boulder, CO: Westview Press, 1994, pp. 279–94.

MacDonald, Stephen, and John Lee. "Assessing the International Competitiveness of the United States Food Sector." In Bredahl, Abbott, and Reed (Eds.), *Competitiveness in International Food Markets.* Boulder, CO: Westview Press, 1994, pp. 191–202.

Miner, William. "Assessing the Competitiveness of the Canadian Food Sector." In Bredahl, Abbott, and Reed (Eds.), *Competitiveness in International Food Markets.* Boulder, CO: Westview Press, 1994, pp. 231–40.

Porter, Michael. *The Competitive Advantage of Nations.* New York: Free Press, 1990.

Reed, Michael, and Mary Marchant. "The Global Competitiveness of the U.S. Food Processing Industry." *Northeastern Journal of Agricultural and Resource Economics* 22 (1992): 61–70.

Vernon, Raymond. "International Investment and International Trade in the Product Cycle." *Quarterly Journal of Economics* 80 (1966): 190–207.

Walter-Jorgensen, Aage. "Assessing the International Competitiveness of the Danish Food Sector." In Bredahl, Abbott, and Reed (Eds.), *Competitiveness in International Food Markets.* Boulder, CO: Westview Press, 1994, pp. 295–308.

Chapter 13

International Marketing: Analyzing Opportunities

The term *international marketing* is used here to portray the firm-level efforts that result in export sales for particular companies. It is the micro-level aspects of trade that result in a flow of products and services to foreign consumers. Exporting and international marketing are used synonymously. For our purposes it is best to envision the firm as a food manufacturer that is considering selling its products in foreign countries. The firm is not simply exporting, because international marketing assumes a conscious effort to understand all aspects of the foreign country to maximize the outcome from the firm's marketing effort. This might mean that the product needs to be changed, new marketing strategies developed, and new personnel/talents added to the firm.

Thus far, we have discussed the macro picture for international trade: policies, exchange rates, preferential trade agreements, the GATT, and foreign direct investment. This chapter covers some of the analyses that firms undertake to improve their likelihood of success in international markets. Certainly the firm must know about the country's market conditions, industry conditions, marketing institutions, and legal restrictions before the firm can successfully market its products.

As the previous chapter on competitiveness indicates, there must be some excellent reasons that a firm would want to overcome all the problems associated with exporting or marketing internationally. Yet exporting is an increasingly common occurrence. The globalization of the world economy has forced firms to, at a minimum, think internationally because they compete with global firms in their home market. Many firms find it profitable to export too.

There are many reasons that encourage a firm to overcome problems with exporting. Exporting allows the firm to increase volume and therefore spread its fixed costs over more units. This improves firm competitiveness and could result in lower costs for domestic and international markets. The home country market might be stagnant or declining, while international markets offer more growth potential. This is particularly true of U.S. food manufacturers, who face saturated domestic markets but vibrant international markets. Firms can extend product cycles and trademarks through international markets if the firms have leading-edge products that are not readily available in other countries. The firm's product can still be sold in international markets when new products have eroded the product's market share in the home country. The firm can also learn from its activities abroad. Through its international operations the firm can observe new products, new ways of doing business, and consumers who view the world differently. These experiences can help the firm compete in all markets.

Often the first time that the firm's product reaches an international audience is less formal, through a broker or wholesaler. Someone will approach the firm with an order that results in the product being exported. The manufacturing firm sells the product at the factory, but the product ultimately is shipped outside the country. Another possibility is that a foreign buyer will approach the company with an order from an international client. This initial dabbling with international markets by the manufacturer might lead the firm into investigating a more studied approach to international markets, which is the subject of this chapter and the next.

The firm's entry into an international market should be the result of a process of analysis and decision making so that the firm can make the most from its efforts. International marketing is similar to domestic marketing because all the main principles still apply: the firm must supply a product that is desired by consumers at a price that is within reach of its customers, the product must be placed in the market such that consumers can easily purchase it, and the product must be promoted and advertised so that consumers know the product's attributes and value.

Yet the challenge of international marketing is that the commercial, economic, and cultural settings differ in each country. The firm must adapt to different languages, religions, customs, rules and regulations, and risks for every country where it chooses to export. The firm must reconsider the product or products that it sells, needed changes that could be made to meet new markets, who uses the product and where, and infrastructure needs associated with the product (e.g., refrigeration).

These challenges are best tackled after an in-depth assessment by the firm and a clear strategy by product and country. Yet one must realize that international marketing is a learning experience, so initial commitments should be made with that in mind. Firms normally begin their international marketing through exporting, but this can progress to direct investment as the firm's expertise and market size increase. The analyses needed are classified as economic, political/legal, cultural, and financial, and are identical for all entry modes. These analyses should be performed for every country that the firm considers, but the firm will normally perform an assessment or initial screening of markets to begin the process.

MARKET ASSESSMENTS

The market assessment begins with macroeconomic variables that indicate whether a potential market for the product exists. Such variables would include GDP per capita, population, a stable government, and a free-market system. These variables

A Model of U.S. Agricultural Export Growth

In their aggregate analyses of U.S. agricultural exports, Salvacruz and Reed used a regression model to predict fast-growing markets. They found that the country's aggregate agricultural imports, GDP growth rate, and self-sufficiency ratio (the country's production divided by its consumption) positively influenced U.S. agricultural exports, while its distance from the United States negatively influenced U.S. agricultural exports. They went on to predict the fastest growing markets for the 1990s in terms of percentage growth and dollar value.

would exclude many countries from consideration because of the product's price, market size, and political considerations. A second stage might involve collection of data that would influence market size, such as the number of refrigerators, microwave ovens, automobile ownership, and household floor space. A third stage would investigate the competitive environment in the foreign market and assess the costs of entry and profit potential. Finally, the firm would establish an ordered list of potential target countries that are candidates for further in-depth study.

There are many types of targeting models that firms can use to make these initial assessments. Walvoord has an elaborate process that takes firms through a general market potential filter (which includes basic economic data, social structures, and geographic features), a filter related to the general market for the product (cultural acceptability, trends in similar products, market size, and taxes), micro-level research associated with the specific product (competition, sales projections, entry costs, and profit potential), and finally corporate factors that might influence the targeted markets. Each of the steps requires more detailed information and data on the project.

IN-DEPTH ANALYSES

After a few targeted countries are identified, in-depth analysis is required to determine more details on the potential market for the product. These analyses cover economic, legal and regulatory, political, cultural, risk, and financial areas. The investigations will provide the information needed to make a final assessment of the product's potential and the background needed to establish a strategy to enter the market successfully.

Economic Analysis

The economic analysis provides the background needed to determine that consumers will purchase the product (in its current or a modified form) at a price that

It Is Not Just Per Capita Income . . .

One cannot simply judge a country based on its per capita income when assessing potential market size: population and the relative size of the middle- and upper-income class may be more important. Three countries—Brazil, India, and Mexico—provide good examples where there is a significant middle- and upper-income class that could purchase imported food products.

In 1998, per capita gross domestic product was $4,709 in Brazil, $3,438 in Mexico, and only $425 in India. The average family cannot afford high-priced imported food products. However, U.S. exporters do not need to sell to every family in these countries to have a profitable entry. If 20 percent of the population in Brazil and Mexico and 5 percent of the population in India are reachable by the exporter, the market sizes would be 33 million (Brazil), 19 million (Mexico), and 49 million (India). These are huge numbers, and they showcase the importance of segmenting the market and determining where the firm should focus efforts.

will allow the firm to have a positive return. Some of the considerations in this analysis come from the initial market assessment, but the effort put into this analysis should be more extensive and detailed.

The country's current standard of living and the distribution of income should be a major concern. Low-income people concentrate their purchases on staple food items, which tend to be grown locally or imported in bulk form (grains and pulses). They do not have much discretionary income to use on highly processed foods. If you are selling highly processed carrots and the country is quite poor, there may be little potential for your product. As incomes increase, people tend to upgrade their diet and consume higher-quality products such as meat, dairy, and eggs. There is more room for variety in the diet, and imported products are more affordable. The income elasticity of demand for food is still high as people's incomes move from low- to middle-income levels (up to $10,000 per capita).

The distribution of income and the number of people in various income classes are very important because firms normally focus their marketing efforts and try to reach certain segments of the population. The overall income of a country is not as important as the number of people within certain income classes. Brazil, India, and Mexico are three countries that are considered poor, but their populations are large and their income distribution is such that there is a large middle class in each country. A firm selling processed foods that cater to a middle-class person might find a large market for the product in these countries.

Ability to purchase is important, but buyer motivation must be understood if the firm is to succeed in selling products. Consumers throughout the world are flooded with products and must choose the ones that best fit their needs. A better understanding behind the motivations to purchase is crucial if a product is to succeed.

Many of the motivations behind purchasing behavior are cultural in nature and, as such, more difficult to quantify. Family structure (i.e., who does the shopping), religion (which products are viewed positively and which are not), and educational systems play a role in buyer motivation.

The structure of the food retailing industry is also important in relation to purchasing decisions and entry strategies. The characteristics of people who purchase food items from warehouse-style stores might be very different from the people who purchase food items from small, local specialty food shops. There are likely wide variations in the nature of those shoppers between countries too. In Hong Kong, most people (male and female) purchase foods from wet markets (small vendors in centralized markets) on their way home from work, but women in France are more likely to shop at such markets during specific shopping excursions. Higher-income people will tend to purchase food from supermarkets and other large enclosed stores that carry many different food items.

The economic analysis should also investigate consumption patterns in detail (by demographic group) and the correlation in consumption among goods. Understanding food consumption habits is crucial for food product companies, and the trends in purchasing behavior are even more important. Many countries are following the Americanized trend in convenience foods, so there are natural opportunities for U.S. food processors to lead in those products if the food item is placed properly in the distribution system. However, the residents of many countries place a higher premium on fresh products than Americans, which inhibits U.S. exports of processed foods.

Transportation systems, information systems, and other infrastructure can have a pronounced influence on consumer purchases. Marketing and promotion are contingent on the types of media available to the firm (television, radio, billboards, etc.), and the phone system will determine the potential access to the Internet, an increasingly important source of information and product supply. The American transportation system is based on the premise that nearly everyone has access to a car. Therefore, much of the shopping is located where retail space is less expensive and parking is readily available (the suburbs). This assumption does not hold in most foreign countries. Cars are less available, and mass transportation systems are more developed. People either shop close to home or use a subway or bus to go to the grocery store. This has ramifications on the type of store frequented by the typical consumer.

Finally, there can be tax benefits to exporting products as a result of duty drawback provisions. A firm that exports a "similar" product can receive a refund of 99 percent of the import duties that it pays for raw materials. Cigarette manufacturers have an incentive to export because they receive a refund on the import duty they paid on imported raw tobacco, saving them money. Firms can also use transfer prices for exports between their subsidiaries to reduce income in high-tax countries and increase income in low-tax countries.

Legal and Regulatory Analysis

There are a host of considerations relating to the laws and regulations that must be studied by a firm before it decides to launch a product in a foreign country. Each

Transfer Pricing

Assume a multinational firm has production facilities for making shirts in Haiti, Mexico, and the United States, and it is required to pay taxes in all three countries based on its income. Further assume that the firm performs initial cutting and sewing in Haiti; ships the product to its facility in Mexico for finer stitching and attaching buttons; and ships the nearly final product to the United States for final refinements, packaging, and sale. The multinational firm will need to charge a price for the product as it moves from country to country. If corporate income taxes are high in the United States and low in Haiti and Mexico, the firm will charge a relatively high price for the shirt (likely much higher than the production costs) to its American subsidiary so that the firm is making more profit in Haiti and Mexico. This will lower its total tax bill because of the lower rates in Haiti and Mexico. These transfer prices can make a big difference in the firm's tax liabilities and its global profit.

country will have its own regulations concerning product labeling, prohibited ingredients, packaging, and specification of weight. This will likely mean that the food product package will need to be redesigned because the U.S. system of weights and measures is not consistent with the rest of the world (the United States deals in pounds, while almost everyone else uses kilograms). Labels might need to include information in another language, though some countries allow foreign language labels to be stuck on containers.

Foreign countries often want food products from the United States to be certified by the U.S. Department of Agriculture or a similar government agency. Normally, an explanation and letter from the USDA stating that the product has met U.S. standards is enough for the foreign purchaser and the customs agent. Other times there are certification bodies or more specific standards that must be met by imports. If this is required, U.S. exporters will need to go through those certification steps to ensure that the product can be sold in the foreign country. Since 1985 the European Union has had a Third Country Meat Directive dictating that all slaughter plants sending meat to E.U. countries must meet E.U. specifications. No foreign plant can export meat to the European Union without certification, by an E.U. meat inspector, that the plant meets the exact standards of the directive.

Food standards in Japan are set by the Ministry of Health and Welfare. It requires that food product labels clearly indicate additives, preservatives, coloring material, spices or flavorings, and the like, and suggests that a certificate with detailed descriptions of the ingredients (i.e., the names of chemical compounds, chemical names and international index numbers on the colors) be attached to each shipment in order to expedite import procedures. It is advised that exporters shipping a new product to Japan have it tested at a laboratory certified by the Ministry of Health and Welfare. These procedures make it more likely that a product will pass easily through Japanese customs.

Firms considering exporting products to a country should investigate general product laws and regulations concerning competitive behavior, liability (especially product liability), bankruptcy, agency relationships, patents, and trademarks. These laws and regulations vary widely by country, and often legal "culture" plays a significant role in determining how rules are applied. If a product is being sold in a foreign country and a product lawsuit is filed, the case will be tried in the foreign country, so the exporting firm's case will not be tried under its home country law. In many countries there is a tendency for the courts and regulators to side with their own people against the foreign exporter.

Sometimes U.S. laws will dictate behavior of companies in foreign markets. The Foreign Corrupt Practices Act of 1977 makes it illegal for any U.S. company to bribe a foreign government official. This law sounds rather innocuous except that bribery is a common practice in some countries (it is part of the culture), and without a bribe there is no way to enter a market or to receive a contract for a service. This effectively shuts U.S. companies out of certain markets unless they can find a way around that regulation. The Helms-Burton Act of 1996 not only prohibits U.S. firms from transacting business with Cuban companies but also makes it illegal for U.S. firms to transact business with third countries that have operations in Cuba. The legality of this law is currently under question, but it is an example where U.S. laws can impinge upon foreign companies.

Foreign governments are often much more involved in regulating and operating commercial enterprises than the U.S. government. A U.S. food firm could find itself competing with a food processor that is owned and operated by a foreign government. This occurs less than in the past because of the recent privatizations throughout the world. Deregulation has also spread throughout the world, which has reduced some of the legal barriers that affect U.S. firms in foreign countries. Entry barriers have been reduced, and price fixing by the government has become less common.

Political Analysis

A political analysis can assist the firm in deciding whether the country will provide a good market for the product in the future. Sometimes the relations with the home country of the exporter will be an important consideration for the other analyses, since politics can change the economic, legal, and cultural settings. Bureaucrats can choose to administer laws and regulations differently depending on the circumstance. If those circumstances include the country of origin for a product, then politics have entered the scene. Japanese bureaucrats have been known to further political objectives by dictating which products should be imported and which should not. Recall the debate on whether opening up the Japanese beef market would result in higher U.S. beef market shares (see Chapter 3).

The Japanese would normally encourage their businesses to import more products from the United States, but the Chinese often discourage American imports for political reasons. The United States is often haranguing the Chinese to improve their civil rights record or to move their economy toward a free market. The Chinese sometimes try to defuse these comments by putting commercial pressure on the United States. The resulting loss in U.S. exports could soften the American stance on these Chinese political issues. There is no question that the general political climate is very important in providing support for a firm attempting to sell products in a foreign market. If one can get a feel for the objectives of the government and understand how to operate independent of political concerns, the better the chance for success.

Specific trade barriers might discourage importation of some products. Local content laws, subsidies for domestic producers, and other nontariff barriers can reduce the market that a U.S. exporter could achieve. Some of these measures may not be legal under the current World Trade Organization (because they are not giving imported products national treatment), and it is possible for the firm to file a complaint. However, in the meantime, the firm may effectively be shut out of the market. It is best to recognize those potential problems before commitments are made.

Cultural Analysis

Culture is a difficult concept to define. It includes the heritage, traditions, and habits of the people. Interlocked in culture is language, religion, family relations, and social relationships. Understanding the culture helps one envision how people will interpret and respond to external forces. If a firm is attempting to convince a potential purchaser that a product is of high quality and can fit into that person's lifestyle, a basic understanding of how the potential purchaser views life is fundamentally important. It is also important to communicate effectively with partners (agents, distributors, retailers, and others), and an understanding of the culture will improve these communications.

Individual purchasing decisions are affected by the values that people place on goods and services. The way that one chooses to spend time and money is intimately related to what one views as worthwhile, which is linked strongly to religious beliefs, family relations, the educational system, and views on work and leisure. These values will influence the types of products that are purchased, but there are other factors that play a role in determining which product brand is purchased.

Reference groups, the groups with which people within a country identify, can play a strong role in determining purchasing behavior. Identifying the types of role models that people relate to can help in assessing the potential for some products. Sports drinks are often effectively promoted if the product can be associated with athletic events in the United States or other countries. One can also take advantage of the target market's image of the exporter's home country.

Many societies place a high value on conforming to established norms, so understanding those norms is essential before entering the market. Fortunately for many U.S. companies, consumers throughout the world are changing their tastes toward Western goods, including convenience foods and Western-style restaurants (fast food and steak houses). This increases the appeal of products from the United States.

In dealing with foreign partners, one must recognize the role of the individual in the society and the power relationships that stem from the group. These concepts have considerable ramifications in understanding corporate governance in different countries. This recognition is very important for Japan because its culture is quite different from that of the United States. Individuality is not encouraged in Japan, so one should not force a Japanese partner into a quick decision. Quick decisions are based on the ideas of one or two individuals and are not characteristic of decisions within the Japanese culture. Such decision making is a result of the highly risk-averse nature of most Japanese. The pain of a failure is much greater than the joy from a success (the exact opposite of the values in American society).

Financial Analysis

Since the firm is ultimately interested in a successful export marketing plan, one must judge the financial aspects of international marketing and analyze the entry and the variable costs of placing the product in the country. Additional staff, payments to foreign distributors or sales agents, logistical costs, and so forth, must be factored into the calculations. Cash flow and rate of return analysis will be useful in establishing baselines for judging profitability. The firm must also estimate the amount of net working capital needed to allow the product to flow smoothly and include the cost of that working capital in the formulas. It is wise to include sensitivity analysis in the calculations to determine how changes in some cost categories influence profitability.

Risk Analysis

Because so many factors are different between domestic and international marketing, there is a need to assess the risks involved and the ways to reduce those risks. There are political risks (from embargoes or nationalization of operations), business risks (the competition may be too strong, partners may be unreliable, property loss may occur), and financial risks (exchange rates may change your costs, or cash flow might not meet projections).

There are ways to reduce some of the risks involved in international marketing. Exchange rate risks can be reduced if the product's price is quoted in the exporter's home country currency or if the exporter hedges sales using forward foreign exchange markets. However, both of these risk-reduction methods involve costs: quoting prices in the home country currency will force the importer to bear the risk of fluctuating exchange rates, and hedging export sales will cost money. Firms can assess their potential foreign partners through many information systems and also use their bank as a source of information on a potential partner's creditworthiness.

Dealing with experienced export service providers will reduce risk too. Exporters should deal with a reputable bank that has correspondent relationships with major international banks. They can sell their product through a confirmed, irrevocable letter of credit (this allows the bank to pay the exporter as soon as the conditions of payment are met). They can deal with a freight forwarder who knows transportation systems, import regulations, cargo insurance provisions, and other logistical matters. Finally, the exporter can purchase political risk insurance through various agencies, such as the Overseas Private Insurance Corporation, a U.S. government organization.

SOURCES OF INFORMATION AND PROGRAMS TO ASSIST EXPORTERS

The U.S. government has many sources of information that can be used in the analyses explained earlier. For agricultural goods, the U.S. Department of Agriculture's Foreign Agricultural Service (FAS) has a wide array of programs to assist exporters. For commercial products (including many processed foods), the U.S. Department of Commerce has many programs. Other government agencies with export information and programs are the Economic Research Service (USDA)—which has a great deal of data and analysis on U.S. exports, foreign production, and other trade-related issues—and the Small Business Administration—which has some export financing programs. For most agricultural and processed food exporters, the FAS will be the best source of information.

The FAS has attachés stationed in almost every country of the world (the Department of Commerce has Foreign Commercial Service employees in almost every country too). The attachés are responsible for assisting exporters, supplying data, and monitoring trade policy disputes. Some of their reports are accessible through the world wide web. The FAS has information on import requirements by country, trade leads (leads from foreign buyers seeking to purchase U.S. agricultural products), and foreign buyer lists. These sources can help exporters determine whether markets exist for their products. The FAS has a Trade Show program that assists U.S. firms in displaying products at an international trade show. Finally, the FAS is the organization that coordinates the export enhancement program (a targeted export subsidy program for U.S. agricultural products) and the General Sales Manager export credit programs.

There are a host of private organizations that also provide useful information to exporters. The World Trade Center network has many sources of information about markets and individual foreign companies, much of it electronically based. The Chamber of Commerce network can be used to obtain information on business statistics and trends. Often it is an excellent source of information on specific local conditions. Commodity trade associations (such as the U.S. Meat Export Federation) can be a valuable resource for potential exporters. They have offices throughout the world for data collection, market research, and commodity promotion. There are hundreds of private companies that have market research reports that could be appropriate for a product, giving customer information, market measurements (size, potential sales), and the competitive, economic, political, and legal environments.

State and local governments are devoting more time and energy to international trade and relations. They often have economic development offices that provide support for potential exporters, and there are trade offices in some foreign countries. Taking advantage of these initial contacts can provide an introduction that is very beneficial. Some economic development organizations will arrange trade missions that are focused on countries, products, or both, to assist local firms in establishing contacts and commercial relations. These trade missions provide a nice way for a firm to make personal contacts and get a feel for the country's market situation. Business and local libraries will have plenty of resource materials on international markets.

Finally, there are international organizations with data readily available for exporters. The Food and Agriculture Organization of the United Nations has production, consumption, and trade data for all UN-member countries of the world. The International Monetary Fund also tracks and publishes macroeconomic data for UN-member countries.

SUMMARY

1. Economic analyses investigate the ability of consumers in foreign countries to purchase the product. The analyses also study distribution systems, consumption patterns, and transportation and other infrastructure.
2. Legal and regulatory analysis investigates the country's regulations concerning product labeling, prohibited ingredients, packaging, and specification of weight. Other important considerations are general product laws and regulations concerning competitive behavior, liability, bankruptcy, agency relationships, patents, and trademarks.
3. Political analysis considers the relationship between the exporting and importing country. This relationship can be very important because bureaucrats can have great influence on how laws and regulations are administered.
4. Cultural analysis investigates the way people in the country react to external forces. A firm must understand how its potential customers view life and the world if the firm is to be successful in its marketing efforts.
5. Financial analysis estimates the cash flow, rate of return, and working capital requirements for the international marketing effort.
6. Risk analysis assesses the risks faced by the exporting firm and examines ways to reduce the risk through various strategies and government programs.
7. There are many sources of export information available to the firm. The U.S. Department of Agriculture's Foreign Agricultural Service has information and programs to assist U.S. agricultural exporters. The Economic Research Service (USDA) has considerable data and analysis on international trade and production levels.

QUESTIONS

1. The world is obviously changing rapidly because of new technologies and communication systems. Will this lead to more U.S. exports of processed foods? Why?
2. Many foreign food manufacturers are having great success in the United States by selling products from their countries. What strategies are some of them using, and how could those strategies be used by U.S. exporters?

3. Mexico is a very important market for U.S. processed foods. Consider how different Mexico is culturally than the United States and how those differences can be overcome by U.S. exporters.
4. Assume that you are exporting popcorn to Korea. Perform an economic analysis to see whether the product has potential.

WEB SITES FOR MORE INFORMATION

Economic Research Service *www.econ.ag.gov*
Foreign Agricultural Service *www.fas.usda.gov*
Food and Agriculture Organization *www.fao.org*
Department of Commerce *www.doc.gov*
Small Business Administration *www.sba.gov*
International Monetary Fund *www.imf.org*

REFERENCES

Salvacruz, Joseph, and Michael Reed. "Identifying the Best Market Prospects for U.S. Agricultural Exports." *Agribusiness: An International Journal* 9, No. 1 (1993): 29–41.
Walvoord, R. Wayne. "Export Market Research." *Global Trade Magazine,* May 1980.

Chapter 14

International Marketing: Developing an International Strategy

The previous chapter covered the types of analyses that must be performed before entering an international market. This background is crucial for the firm to visualize a strategy for its particular product(s). One of the key strategic issues to be addressed by the firm is the extent that a standardized plan can be developed that will be useful for all countries and whether that plan can be a small adaptation of the domestic plan. A standard plan will be more cost-effective because the fixed resources will be stretched further, whereas widely varying plans by country will be more costly and involve more resources.

This chapter covers elements of a strategic plan that will guide the firm's international marketing efforts. The culmination of the planning process is the export marketing plan. The chapter begins with micro-level observations and decisions that the firm must make to enter an international market. The topics include distribution systems, entry strategies, pricing, logistics, and control and decision making. The final part of the chapter summarizes the entire process by reviewing the components of an export marketing plan.

DISTRIBUTION SYSTEMS

One of the most important decisions that the firm must make is what segment of the market it wishes to pursue. Markets are often segmented by the ultimate product user and the type of outlet where the user purchases the product. For frozen potatoes, one segment would involve individual consumers purchasing the product at supermarkets. Another segment would involve restaurants purchasing the product through a distributor. Still another segment would involve the country's army purchasing the product through a centralized government bidding process. Each of these segments calls for different marketing strategies, partnerships with foreign firms, product qualities, pricing, and volumes. The firm must decide which segments fit its organization and product. This combination of market segment and the means to get the product to the user is called the *distribution channel.*

If the firm wishes to target retail consumers, there are numerous ways to approach them: through supermarkets, warehouse stores, department stores, small grocery stores, and vendors (wet markets). Each of these outlets receives its products from different types of suppliers, and the exporting firm must align itself with one or more of those suppliers. Department stores in East Asia have huge grocery sections and often purchase their foreign products directly from manufacturers and import items directly; many warehouse stores handle their import purchases in the same manner. They also deal in large volumes and want substantial price discounts. A firm aligning itself with a distributor in the foreign country may not gain access to these large-volume customers.

Supermarkets and smaller-scale retailers will purchase their products from local distributors, some supplying specialty products while others providing large product lines. Alignment with distributors with larger product lines might give access to a wider variety of retail outlets. Institutional purchasers—such as hotels, hospitals, and restaurants—will purchase through distributors, but often those distributors tend to specialize in institutional foods. A firm might want to choose a distributor that concentrates on institutional foods if that is the segment that is targeted. Price is less of a concern in institutional markets, but volumes may not be as large either.

The Foreign Agricultural Service (FAS) attaché stationed in the target country can provide a wealth of information, contacts, and advice that will help with understanding the distribution system. These USDA employees are observing the market constantly, and they know local market researchers and many of the marketing firms. They also know many of the regulations and channels associated with particular products and can help with specific questions posed by the U.S. firm. Their reports are aimed at assisting the export operations of U.S. food processors. They are a resource that should not be overlooked in choosing the distribution channel.

All the above discussion, though, must be taken with a grain of salt because it is not country-specific, and many times what is true about one country will not hold for another. Further, marketing relationships change over time. The firm must constantly monitor the situation and become familiar with the markets that it chooses. Often, the firm will choose one distribution strategy initially and either change or add other channels to its marketing effort. This is natural and accepted as long as the firm has not exclusively aligned itself with a particular partner in the importing country. More will be said on this topic in the next section.

ENTRY STRATEGIES

Once the market segment is chosen and the distribution channel visualized, the firm must choose the method to get the product to the ultimate user. This is the second important aspect of choosing the distribution channel. This section highlights some of the entry strategies that are possible, including partners and ownership relationships. As will be seen, some entry strategies are negated by the choice of market segment.

The easiest entry strategy is to indirectly export the product. An indirect exporter sells the product before it leaves the home country (the direct exporter is the firm that has title or ownership of the product as it leaves the home country). The firm can sell its product to a domestic buyer, foreign buyer, or an export management company. Indirect exporting is easy, but the processor normally has little control over the product's marketing program after it is sold. All the market placement, promotion, research, and information is lost to the processor. Thus, the processor will have little "feel" for the international market and will surrender control over all future market growth prospects to others. This is a good strategy for some firms because there is little invested in international marketing and no new resources are needed. However, the firm has no control over the future of its international marketing efforts.

A strategy that involves more work for the firm, but more potential payoff, is to directly export the product and sell it to a foreign entity, either a wholesaler, distributor, processor, or retailer. If the sale is to a wholesaler or distributor, the U.S. firm must decide upon the relationship it wants with the foreign marketer, particularly how much involvement it wants in marketing promotion and strategy. The U.S. firm has a vested interest in the foreign marketing firm's activity, so the U.S. firm might want to share in the establishment and funding of the marketing and promotion program. This is something that is negotiated with the foreign marketing firm. If the U.S. firm sells to another processor or retailer, there is usually less concern with the marketing program on the U.S. firm's part because, in the former case, the final product is not identified with the U.S. food processor and, in the latter case, the retailer will have its own marketing program.

The U.S. food processor can also choose to hire its own sales staff in the foreign country or to align itself with an agent in the foreign country. In this case, the U.S. processor keeps title of the product further into the foreign marketing system, but the costs are higher and the risk is likely greater. Nonetheless, the potential return is greater for such an arrangement in many instances because the U.S. firm controls product quality, promotion, and marketing longer and so the placement in the distribution system can be improved. One caution is that if an agency agreement is made with a particular foreign firm and the U.S. firm finds that the arrangement is not generating sufficient returns, it can be quite difficult for the U.S. firm to terminate the agreement. Further, such agreements between U.S. processors and foreign agents are often implied as exclusive commitments, even though they are not specified as such. Thus, if the relationship with the foreign agent does not work out, the U.S. firm might totally lose access to that foreign market.

More advanced entry strategies involve either an ownership interest by the U.S. processor or production in the foreign country. Some of these strategies will not involve exporting, though. The U.S. processor could form a joint venture with a foreign partner. Depending on the structure, the joint venture could involve the U.S. processor exporting to the partner and a jointly owned processing or marketing company. Another entry form would involve full ownership of the foreign facilities by the U.S. processor. The U.S. processor could supply its product as an ingredient for the foreign facility, so exports would be involved in that case. If the U.S. processor licensed its brand or other intellectual property, there would be no exports involved. The U.S. processor would simply be receiving a royalty payment for the intellectual property right (though the U.S. processor would likely have an agreement on production processes, promotion, and other marketing considerations).

A final form of entry that is popular with larger firms is the strategic alliance. In this entry form two or more partners contribute resources to a separate entity, which can take on a corporate form. Each partner contributes unique talents and resources to the alliance, but the alliance is distinct from the individual enterprises of the partners. The alliance can be technology-based, product-based, or distribution-based. A commonly cited example is the European alliance between Nestlé (Switzerland) and General Mills (U.S.) for marketing breakfast cereals. General Mills' production talents were combined with Nestlé's European marketing and distribution talents for a strategic alliance. This particular alliance, however, involves no exports from the United States to Europe.

The choice of a form for market entry depends on factors particular to the firm, product, and country. The competitive structure of the market and the particular market segment will play an important role, as will government rules and regulations. Some forms of entry may not be allowed (such as 100 percent foreign ownership of production facilities). Infrastructure in the market will also impact the entry mode because certain enabling institutions might be necessary to support the firm's strategy. It would be very unusual for a firm to choose a strategy of direct exporting with a sales staff if the firm was new to exporting and marketing structures in the country were undeveloped. Such a focused course by the firm would be very risky.

Ultimately, though, the entry form must be based on an analysis of comparative profit and risk among the alternatives. Firms must be rewarded for their efforts/risks and obtain a return on the capital required in the strategy's implementation. The firm must concern itself with issues of product control, synergy among operations, and potential growth in the future, but profits relative to risk will be the main factor guiding the entry form. The firm must also attempt to maintain flexibility in terms of future marketing options so that growth and changing situations can be easily accommodated.

PRODUCT PRICING

The firm has many choices relative to pricing the product for international customers. These choices can be classified into four categories (similar to categories established by Henneberry): cost-plus, incremental, market, and world. The choice of which category and the specific pricing strategy within the category will be determined by competitive factors, income levels for the targeted group, tax structures in the domestic and foreign country, exchange rates, and financing costs.

Competitive factors might have the largest influence on the choice, since the firm must decide whether to base its pricing on costs or on the prices of competitors. If it is based on costs, cost-plus or incremental strategies must be followed; if it is based on the price of competitors, market or world strategies must be followed.

Full cost-plus pricing would add all the fixed and variable costs involved with the product (domestically and internationally), add some markup, and divide by the amount produced. Costs would vary between domestic sales and export sales, but this strategy does not allow different costs to be factored into prices. Costs involved in exporting are normally higher than domestic costs because of higher transportation costs, tariffs, distribution costs, and the like. (Henneberry provides an export costing worksheet.) If this strategy is upheld strictly, the domestic and foreign prices would be identical. This is not a very good strategy because exports would tend to be priced relatively low (in relation to costs) and domestic sales would be priced relatively high.

A *variable cost-plus* pricing strategy separates per unit costs between markets and adds a markup to arrive at the final price. All fixed and variable costs are included, so it is likely that export prices will be higher than domestic prices. A *marginal cost-plus* strategy takes only variable costs into consideration. If there are scale economies, marginal production costs for exports will be lower, so there is a possibility that such a strategy will result in lower export prices than domestic prices.

Market price strategies result in prices dependent upon the prices of competitors and market conditions, rather than costs. A *market penetration pricing* strategy would involve the firm undercutting competitor prices in order to make sure that foreign customers try the product. Such a strategy normally results in rather low prices relative to costs. A *market share pricing* strategy is when the firm enters the market with a target market share and prices the product so that such a share results. This pricing strategy would result in a price higher than the penetration strategy. A *price skimming* strategy would take advantage of limited supplies and competition, resulting in high prices. Finally, *preemptive pricing* is when a firm lowers its price to combat possible new entrants into a market.

World pricing strategies deal with how prices are determined among a firm's markets. A homogeneous pricing strategy is where the firm charges the same price in all countries. A polycentric pricing strategy allows regional managers to make pricing decisions for all countries in their area. A geocentric pricing strategy is where prices are determined in the home office, but vary by country (allowing cross-subsidization if competition is especially fierce in some locations).

FINDING BUYERS, AGENTS, AND DISTRIBUTORS

After all the information about markets is collected and digested, a strategy for entering the distribution system is developed, and a pricing strategy is decided upon, the firm still needs to find someone or some company to purchase its product. All the research and analysis can be useful only if good customers are located. There are many organizations and resources that can help locate buyers.

The Foreign Agricultural Service's (USDA) attachés and the Foreign Commercial Service's (DOC) officers can help with this process. They have lists of companies within the distribution system that are looking for new products. There are numerous directories and web sites that can provide lists of companies also. It is helpful to attend a trade show to meet potential partners and to get a firsthand view of the market. There are also companies and consultants that specialize in finding foreign partners for firms interested in entering a particular market.

The financial worthiness of these partners can be evaluated through private services (Dunn and Bradstreet, for instance) and banks. Yet a financial analysis will not help in gauging how motivated the partner will be in working for increased sales. This is something that the entering firm will need to judge through interviews and conversations with potential partners. The subject of exclusivity should also be discussed in a forthright manner if the partner is an agent for the entering firm. Most entering firms do not want an exclusive relationship with the partner (meaning that the distributor in the foreign country is the only entity that can sell the food processor's products), but the partner will naturally want such an association. It should be made clear in writing whether an exclusive agreement is in force; otherwise, misunderstandings and legal proceedings could result.

In some less developed countries the need for countertrade and barter will exist. Countertrade occurs when a company can export to a country if it finds a buyer for products from that country. Countertrade is required when the country does not have enough hard currency (currency accepted on the world market) to use for purchasing imports. Barter is a special case where the exporting company is paid through merchandise. In such instances, the company's challenge is to find a buyer for the merchandise from the importing country. In this case, entry, distribution, and pricing strategies are much less relevant because the market in the foreign country is so undeveloped that the firm will want to surrender control of the product immediately upon its arrival in the foreign country. Yet there are excellent opportunities in some countries if the exporting firm can find another firm to buy the bartered or countertraded goods.

PROMOTION STRATEGIES

Once the target countries and partners are chosen, the firm must work with the partners to develop promotion strategies to ensure that the product launch is successful and that the firm's market grows over time. The promotion strategy should recognize and use information on buyer motivations to be effective. If there is a positive view of American products, that should be used in the promotional strategy. If the country has strong nationalist views, then a food processor may want to promote the health-giving aspects of the product.

The firm and partners must also decide on the promotional media. Traditional media sources include radio, television, in-store displays, and direct mailings. Each source will have different expected returns and costs that the firm will need to factor into its decision. Sometimes the food processor will be able to team with other processors for an American food week in major supermarkets

Export Mechanics

It is interesting that the parts of exporting that the new-to-export firm fears most are the least important from a strategic perspective. Many firms fear the export process and all the documentation requirements. This is understandable, and there is no question that the mechanics of exporting are crucial to a successful export program. However, there are many service providers that can handle those mechanics quite easily (for a fee, of course). The more important parts of a strategy are discussed in this chapter. Nonetheless, for completeness, this box will discuss briefly some of the items that must be covered if a shipment is to be successful. The Department of Commerce's *A Basic Guide to Exporting* is an excellent resource for export mechanics.

The terms of the sale, where the transfer of ownership will take place, the price, and any quality characteristics or grading that must be present must be clearly specified. When the product is shipped, there are a host of documents that must be present for smooth delivery. These include the commercial invoice, the bill of lading, the certificate of origin, and the export license. Most goods leaving the United States are covered by a general export license, which does not require a specific application process. An experienced freight forwarder can handle all the transportation, insurance, and documentation requirements for exporting.

Exporting firms are naturally concerned about being paid for their product. There are many different ways of payment, but the most common is the letter of credit (LOC), which is issued to the seller by the buyer's bank. If the LOC is irrevocable, it will be paid even if the buyer is bankrupt (assuming that all conditions of the LOC are fulfilled). If the LOC is confirmed, the promise will be fulfilled even if the buyer's bank defaults. Of course, there is a charge for confirming the LOC, but it is normally worth the money to defray the risk. The exporting firm's bank can help out with all these details, and it should have either an international department or a correspondent relationship with a bank that has one.

and department stores with assistance from the FAS attaché. This allows the firm to not only identify with the United States, but also draw people into the stores to try new products.

GLOBAL CONTROL AND DECISION MAKING

By the time the firm has entered a number of foreign markets, it will have developed a system of control and decision making that works for its product, personnel, and management culture. Communication is often a major stumbling block in large organizations, and communications between the home office and foreign operations are particularly difficult. The global management system must be efficient, responsive, and adaptable if it is to serve the purposes of the firm.

There are three basic approaches that the firm can take in its control and decision making: multidomestic, regional, and global. The multidomestic approach allows most decisions to be made at the local level, with information transmitted back to the home office. Clear authority for decisions resides in the foreign country. The regional approach vests decision making with regional managers (who might be located in the home office or in a foreign location). The global approach has all decisions made by a central international sales manager with control over all operations.

Obviously, the approach taken varies by firm, yet the need for flexible decision making and decentralization of control varies widely depending on corporate culture and product characteristics. If products have been altered tremendously by country, there is much more logic in localized control and decision making. Whereas, if products are identical among countries and the firm exports only from the home country, it is more logical to have a global strategy that links production and marketing among all foreign destinations.

EXPORT MARKETING PLAN—THE CULMINATION OF THE PROCESS

The fulfillment of the company's self-assessment, information gathering, analysis, and strategy with regard to exporting is the export marketing plan. The firm's self-assessment is the natural starting point, where it takes a hard look at its production facilities, marketing programs, financial resources, product logistics, and management/control systems. The bits and pieces from the last two chapters are summarized here to clarify the process.

The export marketing plan should have the following components:

Company and product goals. What are the specific plans for each product and each market? What modifications are needed for the product? How do those modifications fit the specific obstacles and opportunities in the target market? This can be viewed as the summary of the plan because it is based on findings from the components below.

Company and product strengths and weaknesses. What are the positive points the firm can take advantage of in international markets? Why will the products (either in current or adapted form) have an edge over the competition? What are obstacles that must be overcome for success?

Current personnel and needs. What human resources are currently available in the firm that can help with the international effort? Are the present managerial resources capable of handling the additional responsibilities of a sustained international marketing effort? Are there synergistic personnel benefits that can accrue from exporting? What new personnel will be needed to make the international venture successful?

Financial needs. What financial resources will be required for the international venture? These should be classified as long-term and short-term. Can the company acquire these financial resources from existing credit sources? Can the firm take advantage of some government financial programs?

Strategies to follow. What strategies are necessary for the firm to launch into a new foreign market? How might those strategies change as the firm becomes more adapted to the foreign market?

Implementation plan. How can these strategies be incorporated into the present firm structure? How might the emphasis on entering a new foreign market affect home country operations?

SUMMARY

1. Exporters target one segment of the distribution system for their initial entry into a market. The exporter can target food processors, retail consumers, institutional users, and the government.
2. Entry strategies include indirect exporting (selling to some other entity that exports the product), direct exporting to a wholesaler/distributor, direct exporting through a foreign sales staff, producing through a joint venture, producing through direct investment, and a strategic alliance.
3. There are two basic categories for pricing strategies. The first involves pricing based on costs, while the other involves pricing based on competitive factors. Examples of pricing based on costs include full cost-plus, variable cost-plus, and marginal cost-plus. Examples of competitive pricing include market penetration pricing, market share pricing, price skimming, and preemptive pricing.
4. There are a host of resources that can assist the firm in finding buyers or distributors for a product. The most important agency for agricultural products is the Foreign Agricultural Service of the U.S. Department of Agriculture. The U.S. Department of Commerce's Foreign Commercial Service can assist with many processed foods.
5. The export marketing plan is the fulfillment of the firm's self-assessment, information gathering, analysis, and strategy with regard to exporting. The self-assessment forces the firm to analyze its production facilities, marketing programs, financial resources, product logistics, and management/control systems.

QUESTIONS

1. Why do many U.S. exporters begin by selling products in Canada?
2. Popcorn is a food that is very popular in the United States, but much less so in other countries. What strategy would you use if you were a popcorn exporter to increase the market size throughout the world?
3. Why might a firm producing a full line of pickle products want to employ its own sales agent in East Asian countries? What distribution outlet might the firm target?

4. Choose a food item that interests you, and find trade leads and foreign buyer lists for the product. Do you think the product has much potential for exportation based on your findings?

REFERENCES

Henneberry, David. "Pricing Strategies." In C. Parr Rosson (Ed.), *International Marketing for Agribusiness: Concepts and Applications.* College Station, TX: Global Entrepreneurship Management Support, 1992.

Jeannet, Jean-Pierre, and David Hennessey. *Global Marketing Strategies.* Boston: Houghton Mifflin Co., 1998.

Root, Franklin. *Entry Strategies for International Markets.* Lexington, MA: Lexington Books.

U.S. Department of Commerce. International Trade Administration. *A Basic Guide to Exporting.* Government Printing Office, 1994.

Index

173–175, 189–190, 194, 201–202, 210, 214, 216, 222, 224

Transboundary issues, 137, 139, 153, 156

Transfer price, 173, 210–211

Treaty of Rome, 110, 112

Tuna-dolphin dispute, 141–143

U

Unilateralism, 154–156

Unilever, 14, 180

United Kingdom, 7–9, 152, 161–167

United Nations, 5–7, 15, 23, 84, 94, 154, 216

United Nations Conference on Environment, 154

U.S. Department of Commerce, 184, 215, 217, 223–224, 226–227

U.S. Department of Agriculture, 3–5, 10, 15, 90–91, 96, 171, 215–217, 219

APHIS (see Animal Plant Health Inspection Service)

ERS (see Economic Research Service)

FAS (see Foreign Agricultural Service)

GSM (see General Sales Manager program)

Uruguay Round (see GATT)

V

Variable levy, 38, 48, 51, 55–56, 76, 87, 161, 170

W

Welfare, 18–19, 21–22, 25, 31–32, 37–44, 46–49, 50–51, 53–55, 58, 65, 84, 97–99, 102–103, 109, 135, 139, 144–148, 150, 158

Wheat, 8–9, 20, 33–34, 36, 48, 64–66, 91–92, 98–102, 104, 106, 115, 164–166, 169–171

Wine, 11, 16–24, 27–28, 164, 167, 169–170

World Bank (see International Bank for Reconstruction and Development)

World Trade Organization (WTO), 12–13, 69–74, 83–85, 92–93, 96–97, 135, 153, 158, 172, 174, 180, 213